高等职业教育航空装备类专业新形态教材

航空维修工程英语

主　编　田　娟　王超博

副主编　陈跃华　李　娜　皮　芳
　　　　邹　倩

参　编　罗　娜　吴　轮　魏　敏
　　　　江　游　王　海　周召召
　　　　张　鹏　李伟容

主　审　胡　华

北京理工大学出版社
BEIJING INSTITUTE OF TECHNOLOGY PRESS

内容简介

本书对接航空维修岗位群技术领域的日常工作交流英语需求，基于航空器维修人员执照要求、航空维修技术英语等级测试考核的基本特点，融通世界技能大赛、全国职业院校技能大赛、中国技能大赛三大竞赛中英语需求，整合考核知识与技能，重构对应英语职业能力发展的知识内容，涵盖“飞机维修及分类”“服务文件及组织”“典型飞机维修手册”“典型飞机结构”和“典型飞机系统”五大模块 14 个相互关联、能力递进的项目任务，营造真实职业情境。

本书主要面向航空类职业院校飞行器维修工程技术、飞行器维修技术、航空发动机维修技术、飞机结构修理等专业的在校学生，也可以作为航空公司、机场和飞机维修公司航空维修岗位群从业人员的英语专业培训教材。

图书在版编目（CIP）数据

航空维修工程英语 / 田娟，王超博主编. -- 北京：北京理工大学出版社，2023.8
ISBN 978-7-5763-2846-2

Ⅰ. ①航… Ⅱ. ①田… ②王… Ⅲ. ①航空器—维修—英语—高等学校—教材 Ⅳ. ①V267

中国国家版本馆CIP数据核字（2023）第170697号

责任编辑：阎少华　　**文案编辑**：阎少华
责任校对：周瑞红　　**责任印制**：王美丽

出版发行 / 北京理工大学出版社有限责任公司
社　　址 / 北京市丰台区四合庄路6号
邮　　编 / 100070
电　　话 / (010) 68914026（教材售后服务热线）
(010) 68944437（课件资源服务热线）
网　　址 / http://www.bitpress.com.cn

版 印 次 / 2023年8月第1版第1次印刷
印　　刷 / 河北鑫彩博图印刷有限公司
开　　本 / 787 mm × 1092 mm　1/16
印　　张 / 14
字　　数 / 313千字
定　　价 / 55.00元

前　言

党的二十大报告提出："培养造就大批德才兼备的高素质人才，是国家和民族长远发展大计。"随着全球化发展和全球贸易增长，航空维修产业的技术创新和国际化发展对具有英语综合能力的高素质飞机维修人才的需要更加迫切。2020年，中国民航局依照《民用航空器维修人员执照管理规则》发布了《航空维修技术英语等级测试指南》(AC-66-FS-010)，进一步明确了英语相关要求。因此，对具备英语核心素养和岗位核心技能的航空维修高质量应用型人才的培养迫在眉睫。

本书正是在此背景下，探索航空维修工程专业英语"岗课赛证"融通思路，对接航空维修岗位群"会识读，能沟通"英语能力需求，夯实学生语言基础，培养学生学习专业英语和运用专业英语解决飞行器维修岗位群典型任务的能力。通过本书的学习，学习者能提升职场涉外沟通、多元文化交流、语言思维提升、自主学习完善四项英语核心素养。本书力图培养具有中国情怀、航修精神、国际视野，面向世界的航空维修高素质技术技能人才。

本书更新教学理念、重构内容体系、创新数字资源，具有以下特点：

1. 探索"岗课赛证"，重构内容体系

本书对接航空维修岗位群技术领域的日常工作英语交流需求，对接"世界技能大赛""全国职业院校技能大赛""中国技能大赛"三大竞赛中的英语需求，依据民用航空器维修人员执照中《航空维修技术英语等级测试指南》的基本要求，整合考核知识与技能，重构对应英语职业能力发展的五个模块，包含14个相互关联、能力递进的项目任务，构建典型职业能力模块。

2. 对接职业场景，构建语言情境

本书每个情境任务基于航空维修岗位群需求发展逻辑，设计英语语言应用场景。构建"航空情境交流"(职业情境听、说沟通)、"航空维修阅读"(航空维修通识阅读)、"技术文件阅读"(手册等专业素材阅读理解)、"维修工具和设备英语"(航空维修工具识别及使用)、"技术英语规则"(STE等语法规则及要求)、"执照英语实践"(航空维修执照英语题型操练)、"维修安全与规定/代表"(航空维修课程思政)等任务，构建航空维修职业情境。

3. 开发数字资源，培养数字素养

本书配套《航空维修工程英语》精品在线开放课程，以微课资源为核心，开发了课件、音频、视频、动画、虚拟仿真等情境丰富、形式多样、难度递进的数字化学习资源，以二维码形式融入教材，促进学习者利用数字化资源持续学习，发展数字素养。

4. 落实立德树人，弘扬航修精神

航空维修人员具备维修技术相关英语能力不仅是适应我国民航业发展的需要，更与保证飞行安全直接相关。本书关注航空维修作风、安全规定和法则，以我国航空维修产业发展成就、先进人物案例为载体，聚焦敬仰航空、敬重装备、敬畏生命的“三敬”航修精神和零缺陷、无差错的“零无”职业素养，开发航空维修课程思政数字资源，彰显航空特色。

本书校企合作编写团队结构合理、理念先进，由国内航空类高职院校专业骨干教师，实践经验丰富的航空维修、机务维修企业技术专家、全国劳动模范共同组建。本书由长沙航空职业技术学院田娟、深圳航空有限责任公司高级工程师王超博担任主编；长沙航空职业技术学院陈跃华、李娜、皮芳，济南职业技术学院邹倩担任副主编；长沙航空职业技术学院罗娜、吴轮、魏敏，CCAR147 机务培训中心质量经理江游，珠海保税区摩天宇航空发动机维修有限公司全国劳动模范王海，中国人民解放军第五七一九工厂成都航利工程职业教育有限公司周召召，湖北交通职业技术学院张鹏，广州民航职业技术学院李伟容参与编写；广州飞机维修工程有限公司 (GAMECO) 培训总监胡华担任主审。在此，衷心感谢在本书编写过程中给予支持和帮助的校企领导、教师同仁和企业专家。

由于本书遵循全新的编写思路，因此实际编写中难免存在不当和疏漏之处，敬请广大使用者批评指正。期望本书有利于推进我国职业教育英语教学改革，同时为探索教育数字化转型下新型教材的编写提供思路。

智慧职教《航空维修工程英语》在线精品课程网址：https://mooc.icve.com.cn/cms/courseDetails/index.htm?classId=73629c912911983ac0e9cb4d67bd0793。

编　者

目　录 Contents

05 Module 5 Typical Aircraft Systems

Module

01 Aviation Maintenance

Project 1 Aviation Maintenance Development

Objectives

To know the three stages of aviation maintenance development in China.

To understand the general description of fuel servicing.

To master the usage of pounding tools.

To perceive the working style of maintenance.

Part I Aviation Situational Communication

1. Match the call code with its Chinese name and English name.

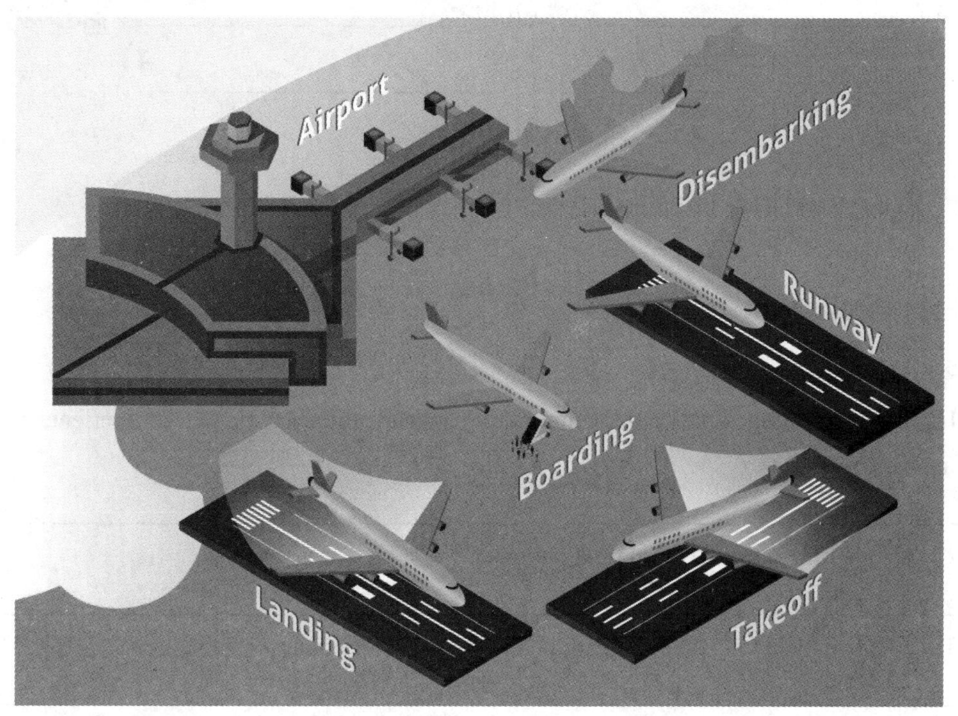

Air traffic control call code

Call code	Chinese name	English name
INFORMATION	区域管制中心	Approach control rader arrivals
DELIVERY	雷达	Approach control radar departure
GROUND	进近管制	Approach control
TOWER	进近雷达进场管制	Radar
DEPARTURE	进近雷达离场管制	Surface movement control
ARRIVAL	塔台管制	Aerodrome control
APPROACH	地面活动管制	Flight information service
RADAR	放行许可	Area control center
CONTROL or CENTER	飞行情报服务	Clearance delivery

2. Listen to some call codes and try to write them down.

(1)____________ (2)____________

(3)____________ (4)____________

(5)____________ (6)____________

(7)____________ (8)____________

Listen

Part II Aviation Reading

2.1 Aviation Reading

Pre-reading Questions

1. How many stages has the development of our aviation maintenance experienced? And what are they?

____

2. Who is "the father of Chinese aviation"?

____

3. When and where is the first maintenance base of our country established?

____

Development of Aviation Maintenance in China

Development of Aviation Maintenance in China

The development of aviation maintenance in China has experienced three stages.The first stage is before the establishment of the People's Republic of China, and we experienced the change from nothing to something. Then from the early years after the establishment of the People's Republic of China to the 1970s, it's the start-up stage. Finally from 1980s to the present, it's an international integration stage.

From nothing to something (Before the establishment of the People's Republic of China)

In 1909, Feng Ru, " the father of Chinese aviation ", made his first airplane and successfully tested it in the United States. He returned to China in 1911 and died in an unfortunate crash during an air show in Guangzhou in August 1912. At that time, the aircraft maker, pilot, maintenance personnel are one person.

In 1912, Sun Yat-sen set up an aviation bureau to establish airports and aircraft factories in Guangzhou, Hangzhou, Kunming, Nanjing and other citys. In 1913, the government of the Republic of China set up an aviation school in Beijing Nanyuan (including a factory to park aircraft and a factory to assemble and repair aircraft).At this time, there are few full-time maintenance personnel, and aircraft maintenance is subordinate to assembly manufacturing. In 1919, the Beiyang government bought 135 aircrafts, but due to financial problems and lack of maintenance, a large number of aircraft parts corroded.At the initial stage of aviation development in China, this is the inevitable consequence of not attaching importance to maintenance and doing nothing if there is no fault.

In 1930, a joint venture between China and the United States, was founded. In 1931, Sino-German joint venture founded Eurasian Airlines, which was reorganized as Central Airlines in 1943.

In this stage, the maintenance system of the United States is adopted, including outfield maintenance and infield repair, which is similar to the current maintenance organization.

Start-up stage (From the early years after the founding of the People's Republic of China to the 1970s)

In 1949, Civil Aviation Administration was established under the People's Revolutionary Military Commission under the direction of the Air Force. Due to the Hong Kong " Two Airlines Uprising ", 12 planes flew back to the mainland, a large number of technical and business personnel became the backbone force in the construction of civil aviation in New China.

In 1950, the Sino-Soviet Civil Aviation Corporation was established in Beijing, which was handed over to the Chinese government in 1954.

Soviet aircraft were used mainly from the 1950s. At that time, maintenance consists of apron maintenance work and infield repair work. The maintenance organization mode was three-level, and mainly adopted hard time maintenance.

International integration stage（From 1980s to the Present）

In 1978, CAAC set up the Aviation Engineering Department, responsible for overall maintenance engineering work. The new aircraft maintenance outline is mainly based on on-condition and condition monitoring maintenance, supplemented by regular maintenance.

In 1980, CAAC established China's first maintenance base in Beijing Capital Airport, which realized the integration of field and internal field maintenance in terms of organizational structure, and realized the integration of production, technology and aviation material.

Since 1987, major airlines have been established. In 1989, Beijing Aircraft Maintenance Engineering Corporation（AMECO, Fig. 1-1）was established as a Sino-German joint venture. Guangzhou Aircraft Maintenance Engineering Company was established（GAMECO, Fig. 1-2）.

Fig. 1-1 AMECO

Fig. 1-2 GAMECO

Civil aviation maintenance market has spawned many maintenance enterprises.

The major companies have adjusted the maintenance organization, adopted new maintenance methods, and gradually realized the integration with international civil aviation.

Words and Expressions

aviation [ˌeɪvɪˈeɪʃən] *n.* 航空；飞机制造业

maintenance [ˈmeɪntɪnəns] *n.* 维修；维护

aircraft [ˈeəˌkrɑːft] *n.* 航空器；飞机

pilot [ˈpaɪlət] *n.* 飞行员；驾驶员

assembly [əˈsemblɪ] *n.* 组装；装配；（待装配的）成套部件，零件组合

personnel [ˌpɜːsəˈnel] *n.* 职员；人事部门

civil [ˈsɪvl] *adj.* 平民的；民用的

apron [ˈeɪprən] *n.* 停机坪

infield [ˈɪnˌfiːld] *n.* 内场

airlines [ˈeəlaɪn] 航空公司

CAAC *abbr.* 中国民用航空局

AMECO *abbr.* 北京飞机维修工程有限公司

GAMECO	*abbr.* 广州飞机维修工程有限公司
hard time maintenance	定时维修
condition monitoring maintenance	状态监控维修
on-condition maintenance	视情维修
regular maintenance	定期维修

Fill in the blanks with the proper words given below, changing the form if necessary.

aircraft	maintenance	personnel	pilot	civil

1. The operator has to be able to carry out routine ________ of the machine.
2. Please remain seated until the _______ has come to a halt.
3. The crash happened seconds after the _______ reported engine trouble.
4. One of the problem areas is lax security for airport ________.
5. After 10 years of rapid development, China's ______ aviation industry is poised to further increase its clout in the world market.

2.2 Manual Reading

There are some tasks for you to fulfill. You should read the Manual Reading materials carefully and finish the tasks.

FUEL- SERVICING

General

(1) This procedure has these tasks:

① Precautions and Limits for the Refuel Operation.

② Prepare the Airplane for a Refuel Operation.

③ Pressure Refuel Procedure.

④ Refuel Operation When the Refuel Quantity Indicators Flash.

⑤ Refuel Operation When the Fuel Quantity Indicating System (FQIS) does Not Operate.

⑥ Pressure Refueling Operation for a Refuel Valve That does Not Open Electrically.

⑦ Fuel System Drainage.

⑧ Drain the fuel From the Sumps after Defueling.

(2) You must not permit the fuel tanks to collect too much water. Do the procedure to drain the sumps drain valves for each tank regularly.

(3) Fuel Servicing Regulations:

① Each operator is responsible for complying with the local, state and national regulations regarding aircraft fuel servicing. It is possible that fire codes and standards make it necessary touse different or more restrictive procedures than those given below. Make sure the procedures used during the refuel operation give sufficient protection to persons and equipment.

② Obey all of the safety precautions supplied in this task: " Precautions and Limits for the Refuel Operation".

③ If you make a decision not to do this recommended procedure, you must have an approved alternative procedure.

Selected from AMM

Words for Reference

refuel	[riːˈfjʊəl]	*vt.* 给……（再）加油；给……续燃料
		vi.（再）加油；续燃料
precaution	[prɪˈkɔːʃn]	*n.* 注意事项
indicator	[ˈɪndɪˌkeɪtə]	*n.* 指示灯
valve	[vælv]	*n.* 阀门，活门
drainage	[ˈdreɪnɪdʒ]	*n.* 排水系统
sump	[sʌmp]	*n.* 滑油罐
tank	[tæŋk]	*n.* 槽，箱
fuel servicing		燃油勤务
FQIS（Fuel Quantity Indicating System）		燃油量指示系统
codes and standards		规范与标准

Reading Tasks

1. Judge the following statements according to the contents of the AMM.

（1）You are banned from collecting too much water when you make refuel operation.（　）

（2）Each operator should not obey all the safety precautions supplied in " Precautions and Limits for the Refuel Operation".（　）

（3）The procedure of refuel operation is the same when the refuel quantity indicators flash and FQIS does not operate.（　）

2. Fill in the blanks according to the above reading materials.

（1）You should _____ the fuel from the sumps after defueling.

（2）You must adopt the ________ procedure to do refuel operation if you don't follow the above recommended procedure.

Translate the following sentences into Chinese.

1. You must not permit the fuel tanks to collect too much water.

____

2. Do the procedure to drain the sumps drain valves for each tank regularly.

____

3. Each operator is responsible for complying with the local, state and national regulations regarding aircraft fuel servicing.

____

4. Make sure the procedures used during the refuel operation give sufficient protection to persons and equipments.

____

5. If you make a decision not to do this recommended procedure, you must have an approved alternative procedure.

____

Translate the following abbreviations into corresponding Chinese.

1. CAAC______________________________
2. AMECO______________________________
3. GAMECO______________________________
4. fuel servicing______________________________
5. FQIS(Fuel Quantity Indicating System)______________________________

Part III Tools and Equipments

Pounding Tools

1. Match the words or phrases with the translations.

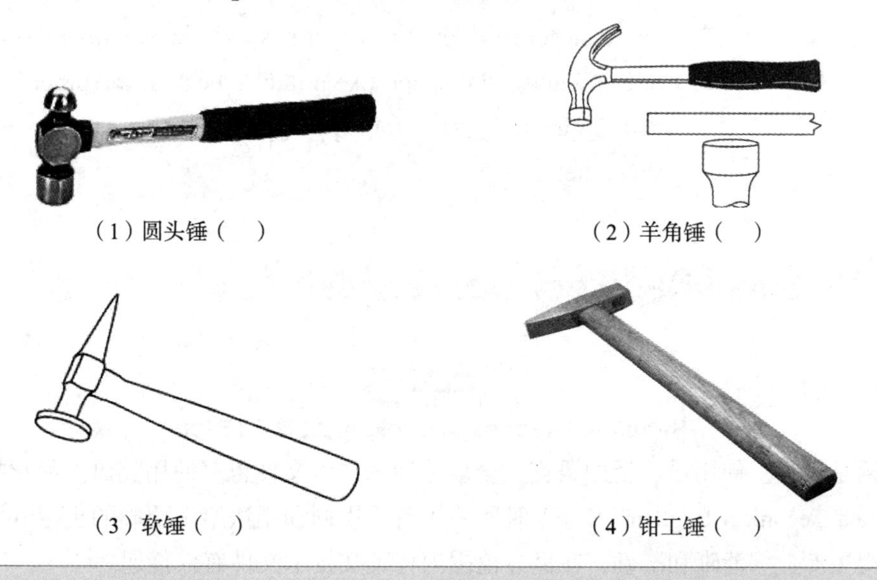

(1) 圆头锤( ) (2) 羊角锤( )

(3) 软锤( ) (4) 钳工锤( )

A. Ball Peen Hammer B. Machinist's Hammer
C. Soft-Face Hammer D. Claw Hammer

2. **Fill in the words or phrases in the blanks.**

Ball Peen Hammer	Claw Hammer
Soft-Face Hammer	Machinist's Hammer

(1) This hammer is used for driving and removing nails, but is seldom used when working on an aircraft. It is not designed for use in metal working because its face is slightly crowned to concentrate the force when driving nails.

(2) This is the most widely used hammer for general aviation maintenance; available with head weights from a few ounces to several pounds. The face of the hammer is flat with slightly rounded edges, and the opposite end of the head is rounded like a ball.

(3) This hammer is used for the forming of soft metals and for tapping operations on surfaces that are prone to damage. The materials used for the hammer head include cowhide, rubber, plastic, and synthetic materials.

(4) This hammer is used to strike objects to make them move or deform. The head of it has two sides, with a flat head at one end and a flat shape like a duck's beak at the other. The flat end is generally used for tapping, and the flat beak is used for sheet metal.

Part IV Simplified Technical English

Simplified Technical English, ASD-STE100

英语是世界上使用最广泛的语言，全球流通的技术文件也多使用英语。简化技术英语（Simplified Technical English，STE）对所使用英语用词和用法给出明确的规定和限制，让表达变得更专业、清晰和准确，在世界范围内将英文技术文件有效传递。

What is STE?

Simplified Technical English (STE) is an international specification for the preparation of technical documentation in a controlled language.

What is the necessity?

In aviation, it is mandatory to correctly understand maintenance and operation documentation to make sure that systems operate safely and correctly and to protect human lives. In the late 1970s, the Association of European Airlines (AEA) asked the European Association of Aerospace Industries to write such documentation. In 2005, the ASD Simplified Technical English Specification, ASD-STE100 published.

What is the purpose?

The purpose of STE is to give technical writers guidelines on how to write technical texts in a clear (清晰), simple (简单) and unambiguous (明确) manner.

Can STE be used alone?

No. It is intended to be used with other applicable specifications for technical publications (技术出版物), style guides, and official directives (官方指令). A high standard of professionalism is necessary to use the STE specification correctly.

Part V Maintenance Licence English Application Practice

Choose the best answer from the following choices.

( ) 1. It includes those maintenance practices necessary to prepare the aircraft for weighting.

A. 它包括了准备飞机顶升时的必要维护工作的信息。

B. 它包括了准备飞机称重时的必要工作的维护信息。

C. 它包括了准备飞机称重时的有用的维护信息。

( ) 2. Any maintenance anomaly would subsequently be brought to the attention of both the owner and maintenance personnel.

A. 任何维修异常都应立即通知营运人和维修人员。

B. 任何维修异常随后都会引起营运人和维修人员的注意。

C. 任何维修异常都应按顺序通知营运人和维修人员。

( ) 3. High-time airplanes, disassembled for major overhaul, were inspected to assist in selecting inspection items.

A. 对于使用寿命长的飞机应进行检查以协助选择检查项目，大修时应进行拆卸。

B. 为了大修而拆卸的及时飞机应被检查以协助选择检查项目。

C. 为了协助选择检查项目，应检查并拆卸使用寿命长的飞机进行大修。

( ) 4. Removal/Installation: The removal/installation portion shall state clearly the sequence of steps required to remove and reinstall a component or unit, along with precautions to be observed.

A. 拆卸 / 安装：拆卸 / 安装部分应清楚说明拆卸和安装部件或组件所需的步骤顺序，以及遵循的预防事项。

B．拆卸 / 安装：拆卸 / 安装部分应清楚说明拆卸和重新安装部件或组件所需的步骤顺序，以及遵循的预防事项。

C．拆卸 / 安装：拆卸 / 安装部分应清楚说明拆卸和重新安装部件或组件以及遵循的预防事项所需的步骤顺序。

（ ）5．“Maintenance Practices” which are a combination of servicing，removal installation，adjustment test，inspection check，cleaning painting，and approved repairs should follow the troubleshooting section.

A．“维修施工”包括维修、拆卸安装、调节测试、检查、清理喷漆或故障排除部分批准的修理。

B．“维修施工”包括勤务、拆卸安装、调节测试、检查、清理喷漆及批准修理，且遵循故障排除部分。

C．“维修施工”包括维修、拆卸重装、调节测试、检查、清理喷漆和故障排除部分批准的修理。

（ ）6．Pull the vehicle to a safe area where you can use the standard procedures.

A．把车拖到安全的地方，在那里可以使用标准程序。

B．拉动车到安全的区域，你可以使用标准过程。

C．把车拖到平坦的地方，在那里可以使用标准程序。

（ ）7．用扳手拧松并拆下螺母。

A．Loosen and remove nut with wrench.

B．Tighten and torque nut with spanner.

C．Remove and replace nut with wrench.

（ ）8．不定期维修检查是指制造商认为特殊或不寻常的情况下对飞机及其系统和组件进行的维修检查和检验，这些检查和溢验与上 −10 规定的时限无关。

A．Unscheduled maintenance are those maintenance checks and inspections on the aircraft，its systems and units which are recognized by the manufacturer as special or unusual conditions which are not related to the time limits specified in −10 above.

B．For the unscheduled maintenance checks and inspections of aircraft，aircraft systems and components，the manufacturer shall not include special or unusual conditions from the time limits specified in −10.

C．If the unscheduled maintenance checks and inspections on the aircraft，its systems and units are special or unusual conditions，the time limits specified in −10 shall be referred to by the manufacturers.

（ ）9．Inspection times will be adjusted for any severe problems.

A．检查时间将根据一些问题进行调整。

B．针对严重问题将调整检查次数。

C．针对严重问题将延长检查时间。

( ) 10．在你继续测试之前，间隔5分钟是必要的。

A．The test can be canceled as long as 5 minutes is left after the text.

B．5 minutes is a must unless you move on the test again.

C．An interval of 5 minutes is necessary before you continue the test.

( ) 11．Because airplanes are intended to have long lives，their flight safety must be assured.

A．In order to make airplanes have long lives，they are often in good condition.

B．To make airplanes have long lives，it is necessary to ensure their flight safety.

C．Because the long lives of airplanes are required，make sure the mechanics do the maintenance task correctly.

( ) 12．These tests should require no special equipment or facilities other than that installed on the aircraft and should be comparable to the tests performed by the flight crews.

A．These tests should only use the equipment or facilities installed on the aircraft.These tests should be able to perform by the flight crews.

B．These tests should not use the special equipment or facilities that are not installed on the aircraft. These tests should be similar to the tests performed by the flight crews.

C．These tests should require the equipment or facilities installed on the aircraft rather than the special ones and should be replaceable by the tests accomplished by the flight crews.

( ) 13．Recheck the assembly filter for leakage.

A．Test the assembly filter for leakage again.

B．Check the assembly filter for leakage again.

C．Re-inspect the assembly filter for fuel leakage.

( ) 14．Landing gear position has no effect on stall speed.

A．Landing gear position will not effect its speed.

B．The stall speed will be effected by landing gear position.

C．The stall speed will not be effected by landing gear position.

( ) 15．If cracks are found, notify the chief inspector.

A．If you find cracks, notify the chief inspector.

B．If cracks happen, tell the chief inspector.

C．If you find cracks, tell the chief inspector.

Part VI Safety and Regulations

Working Style of Maintenance

"We must remember that these are people, not numbers!"

President Xi Jinping put forward eight aspects of style construction requirements: Truth-seeking and pragmatic, focus on subtle, take the initiative to visit, listen to the voice, their own excellent, continuous improvement, problems, strict self-discipline. At the same time, the "three strictness and three authenticity" requirements for style construction: Not only strict in self-cultivation, strict in the use of power, strict in self-discipline, but also practical in planning, entrepreneurship, and life.

China's civil aviation maintenance industry should not only adhere to the guarantee of flight safety as the bottom line, but also take into account the need to ensure the normal flight and reduce costs, and the maintenance style should be further changed to the connotation of "preciseness, professional and integrity".

Preciseness is the basic characteristic and requirement of civil aviation maintenance. Professional means to have professional qualifications as the basis, clear industry role. Integrity is the bottom line that the maintenance industry must comply with.

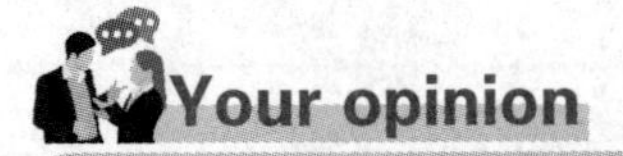

Your opinion

How do you see the development process of aviation maintenance in China? How to do scientific maintenance nowadays?

Project 2 Methods of Aviation Maintenance

Objectives

To know the several methods of aviation maintenance.

To understand the precautions for engine maintenance.

To master the usage of turning tools.

To perceive the safety and speed of development of China's civil aviation.

Part I Aviation Situational Communication

1. Read aloud the topic-related sentences for aviation communication and try to match them with their Chinese meaning.

Airport control common expressions

_______ (1) Say again all before departure frequency.

_______ (2) You are unreadable.

_______ (3) How do you read?

_______ (4) I say again.

_______ (5) Speak slower.

_______ (6) Roger.

_______ (7) Negative.

_______ (8) Contact.

Listen

A. 你听我声音怎样?	B. 离场频率前的内容再说一遍。
C. 收到。	D. 无法听清。
E. 请减慢语速。	F. 与……联系。
G. 不予许可（不正确）。	H. 我重复一遍。

2. Listen to some sentences for aviation communication carefully and try to fill in the blanks.

(1) You are __________.

(2) ________.

(3) ________ do you read?

(4) ________.

(5) Say ________ all before departure frequency.

Listen

Part II Aviation Reading

2.1 Aviation Reading

Pre-reading Questions

1. How many kinds of checks are there according to the passage? And what are they?

____

2. What is HMV?

____

3. On average, how many times D checks does a commercial aircraft undergo before being retired?

____

Maintenance Tasks

Maintenance tasks usually include service and maintenance work, preventive maintenance work, engine maintenance work and parts repair work.

Maintenance work usually includes detection (inspection or testing), repair, modification, (excluding design) renovation, airline maintenance, service, regular maintenance.

For aircraft airline maintenance and inspection maintenance, maintenance task commonly includes lubrication or service, operation inspection, visual inspection, inspection or function inspection, recovery, scrap.

Maintenance steps are normally receiving inspection, contract (work order) confirmation, initial inspection, concealed damage inspection, process inspection, final inspection, release (including retention items), records and reports.

Maintenance elements are generally plant facilities (working environment). Tools and Equipments, maintenance documents and work cards, work procedures and personnel.

ABC Check System

ABC Check System

Aircraft maintenance checks are periodic inspections that have to be done on all commercial and civil aircraft according to certain usage parameters—such as flight hours, calendar time, or flight cycles.

Airlines and airworthiness authorities casually refer to the detailed inspections as " checks ", commonly one of the following: A check, B check, C check, or D check. A and B checks are lighter checks, while C and D checks are considered heavier checks.

A Check

The A check is performed approximately every 400–600 flight hours, or every 200–300 flight, depending on aircraft type. It needs about 50–70 man-hours, and is usually performed in an airport hangar. The A check takes a minimum of 10 man-hours.

B Check

The B check is performed approximately every 6–8 months. It takes about 160–180 man-hours, depending on the aircraft, and is usually completed within 1–3 days at an airport hangar. A similar occurrence schedule applies to the B check as to the A check. B checks are increasingly incorporated into successive A checks.

C Check

The C check is performed approximately every 20–24 months. This maintenance check requires large majority of the aircraft's components to be inspected. The aircraft must not leave the maintenance site until it is completed. This check puts the aircraft out of service for 1–2 weeks. It also requires more space than A and B checks, therefore, it is usually carried out in a hangar at a maintenance base. The effort needed to complete a C check is up to 6,000 man-hours.

D Check

The D check, sometimes known as a " Heavy Maintenance Visit" (HMV), is by far the most comprehensive and demanding check for an airplane. This check occurs approximately every 6–10 years. It is a check that more or less takes the entire airplane apart for inspection and overhaul. Such a check can generally take up to 50,000 man-hours, and 2 months to complete depending on the number of technicians involved. It also requires the most space of all maintenance checks, and as such must be performed at a suitable maintenance base. On average, a commercial aircraft undergoes two or three D checks before being retired.

Words and Expressions

detection	[dɪ'tekʃ(ə)n]	*n.* 察觉，发现；侦破（案件）
inspection	[ɪn'spekʃ(ə)n]	*n.* 视察；检查，审视
modification	[ˌmɒdɪfɪ'keɪʃ(ə)n]	*n.* 修改的行为（过程）；修改，更改；修饰
renovation	[ˌrenə'veɪʃ(ə)n]	*n.* 翻新，整修
service	['sɜːvɪs]	*n.* 公共服务系统，公共事业；服务；接待
lubrication	[ˌluːbrɪ'keɪʃn]	*n.* 润滑；润滑作用
hangar	['hæŋə(r)]	*n.* 飞机库；飞机棚
visual inspection	['vɪʒuəl][ɪn'spekʃn]	目测；目检；外观检验
equipment	[ɪ'kwɪpmənt]	*n.* 设备，用具
document	['dɒkjumənt]	*n.* 文件，公文，文献；证件，单据

procedure	[prə'siːdʒə(r)]	*n.* 手续，步骤
parameter	[pə'ræmɪtə(r)]	*n.* 界限，范围；参数，变量
flight cycle	[flaɪt]['saɪk(ə)l]	飞行周期
component	[kəm'pəʊnənt]	*n.* 组成部分，成分，部件
overhaul	[ˌəʊvə'hɔːl]	*n.* 大修，彻底检修
HMV		*abbr.* 重度维护检查

Fill in the blanks with the proper words given below, changing the form if necessary.

inspection	renovation	service	equipment	component

1. Engineers carried out a thorough _______ of the airplane.
2. New _______ is urgently needed.
3. Trust is a vital _______ in any relationship.
4. The hospital needs _______.
5. This car is overdue for a _______.

2.2 Manual Reading

There are some tasks for you to fulfill. You should read the Manual Reading materials carefully and then finish the tasks.

WARNING: DO NOT TOUCH THE COMPONENTS OF THE OIL SYSTEM IF THE ENGINE IS HOT. THESE COMPONENTS STAY HOTTER THAN OTHER COMPONENTS. HOT COMPONENTS CAN BURN YOU.

WARNING: DO NOT OPEN THE OIL SYSTEM UNTIL THE PRESSURE GOES TO ZERO. THE PRESSURE GOES TO ZERO APPROXIMATELY 5 MINUTES AFTER AN ENGINE SHUTDOWN. A PRESSURIZED ENGINE CAN RELEASE A SPRAY OF HOT OIL THAT CAN BURN YOU.

WARNING: DO NOT LET HOT OIL GET ON YOU. PUT ON GOGGLES AND OTHER EQUIPMENT FOR PROTECTION OR LET THE ENGINE BECOME COOL. HOT OIL CAN BURN YOU.

WARNING: DO NOT LET ENGINE OIL STAY ON YOUR SKIN. USE ENGINE OIL IN AN AREA WITH GOOD VENTILATION. ENGINE OIL IS POISONOUS AND CAN BE ABSORBED THROUGH YOUR SKIN. ENGINE OIL FUMES CAN IRRITATE YOUR RESPIRATORY TRACT.

CAUTION: DO NOT LET HOT OIL GET ON THE ENGINE OR OTHER COMPONENTS. IMMEDIATELY CLEAN THE COMPONENT IF OIL FALLS ON IT. OIL CAN CAUSE DAMAGE TO PAINT AND RUBBER.

CAUTION: DO NOT LET ALKALINE CLEANING FLUID GET INTO THE ENGINE OIL. VERY SMALL QUANTITIES OF THIS FLUID CAN CHANGE THE ENGINE OIL. DAMAGE TO EQUIPMENT COULD OCCUR.

Selected from AMM 202

Words for Reference

component	[kəm'pəʊnənt]	*n.* 组成部分；零件；部件
pressure	['preʃə]	*n.* 压力
approximately	[ə'prɒksɪmɪtlɪ]	*adv.* 大概；大约
shutdown	['ʃʌtˌdaʊn]	*n.* 关闭；停止运行
release	[rɪ'liːs]	*n.* 放出；排放；*vt.* 释放
spray	[spreɪ]	*n.* 喷雾；水沫
goggle	['gɒgl]	*n.* 护目镜
ventilation	[ˌventɪ'leɪʃən]	*n.* 空气流通；通风
irritate	['ɪrɪˌteɪt]	*vt.* 使过敏；使发炎
respiratory	['respɪrətərɪ]	*adj.* 呼吸（器官）的；影响呼吸（器官）的

Reading Tasks

1. Judge the following statements according to the contents of the AMM.

(1) Do not open the oil system until the pressure goes to zero. (   )

(2) The pressure goes to zero approximately 3 minutes after an engine shutdown. (   )

(3) Engine oil fumes can irritate your respiratory tract. (   )

2. Fill in the blanks according to the above reading materials.

(1) Do not _______ the components of the oil system if the engine is hot.

(2) Use engine oil in an area with ________ventilation.

(3) Put on goggles and other equipment for protection or let the engine become cool. Hot oil can burn you.

Translate the following sentences into Chinese.

1. The pressure goes to zero approximately 5 minutes after an engine shutdown.

____

2. Engine oil is poisonous and can be absorbed through your skin.

____

3. Engine oil fumes can irritate your respiratory tract.

____

4. Do not let hot oil get on the engine or other components.

____

5. Very small quantities of this fluid can change the engine oil.

Part III Tools and Equipments

Turning Tools

1. Match the words or phrases with the translations.

（1）开口扳手（　）　（2）梅花扳手（　）

（3）套筒扳手（　）　（4）内六角扳手（　）

（5）组合扳手（　）　（6）扭矩扳手（　）

A. Open-end wrench	B. Box-end wrench
C. Combination wrench	D. Socket wrench
E. Allen wrench	F. Torque wrench

2. Fill in the words or phrases in the blanks.

Open-end wrench	Box-end wrench	Combination wrench
Socket wrench	Allen wrench	Torque wrench

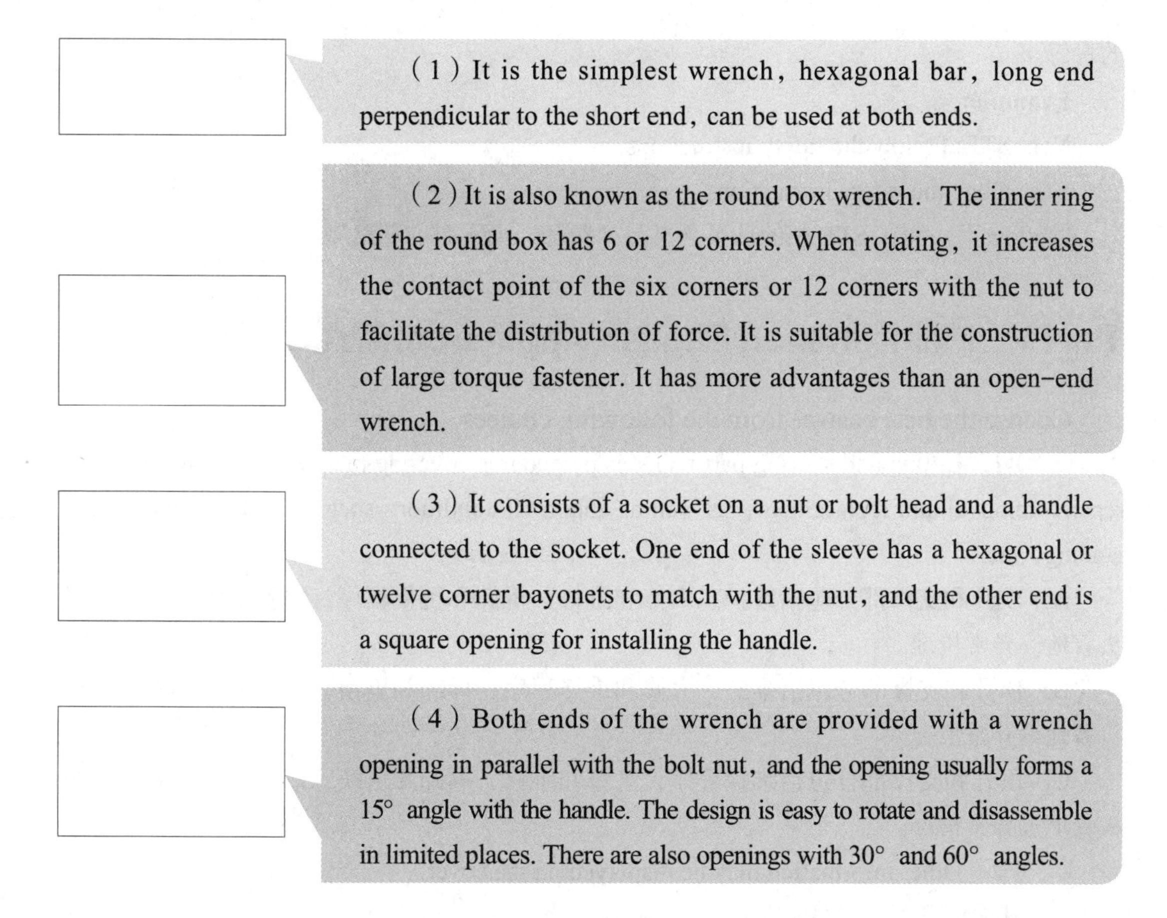

Part IV Simplified Technical English

Part of Speech
词性

★规则：在标准英语中，很多单词有不同的词性，但在 STE 中只能使用规定的词性。

Examples 1: check

The word "check" is an approved noun, but not an approved verb.

Non-STE: Check the oil system.

STE: Do a check of the oil system.

Examples 2: operate

The word "operate" is an approved verb. The word "operable" is an unapproved adjective.

Non-STE: Make sure that the valve is operable.

STE: Make sure that the valve can operate.

Approved Meaning
认可的含义

★规则：需使用认可单词的认可含义。

词典中的每一个认可的词都有一个特定的核准词义，一定要用这些词来表达他们认可的意思。

Example:

Non-STE: Follow the safety instructions.

STE: Obey the safety instructions.

“follow”一词在 STE 中的认可含义是“跟随”而不是“服从”。

Part V Maintenance English Application Practice

Choose the best answer from the following choices.

(   ) 1. Lifting and shoring part includes procedures covering maintenance, overhaul and repair, removal and replacement, as well as abnormal conditions such as belly landing, nose landing, etc.

A. 顶升和支撑部分包括维护、大修和维修，拆卸和更换程序，以及如腹部着地、机头着地等异常情况。

B. 顶升和支撑部分包括维护、大修和维修程序，拆卸和更换，以及如腹部着地、机头着地等异常情况。

C. 顶升和支撑部分包括维护程序，大修和维修，拆卸和更换，以及如腹部着地、机头着地等异常情况。

(   ) 2. Other information may be displayed on the cover.

A. 其他信息可能会显示在封面上。

B. 其他信息可能会在盖板上留下。

C. 其他信息可能会显示在目录上。

(   ) 3. Adjustment/Test: The adjustment test description shall provide instructions, for accomplishment of a test or check to assure the integrity as an operational component of a system subsystem or unit assembly.

A. 调节 / 测试：调节测试说明应提供指示以完成测试或检查及确保系统、子系统或组件中装配组件的完整性。

B. 调节 / 测试：调节测试说明应提供完成测试或检查的指示，以确保作为系统、子系统或组件装配的操作部件的完整性。

C. 调节 / 测试：调节测试说明应提供指示以完成测试或检查及确保作为系统、子系统或组件装配的操作部件的完整性。

(   ) 4. The first step should always be job set up and last should be close up or clear-up.

A. 工作准备总是第一个步骤，收尾或收工是最后一个步骤。

B. 工作准备总是第一个步骤，收尾或收工是接下来的步骤。

C. 第一个步骤有时是工作准备，收尾或收工是接下来的步骤。

(   ) 5. These safety precautions are the minimum necessary for work in the maintenance area.

A. 这些安全措施是在维修区工作的最低要求。

B．这些安全警告满足维修区工作的全部要求。

C．这些安全提示有必要作为维修区工作的要求。

（　）6．Identify the disassembled parts with tags，this method will help you during the subsequent assembly procedures.

A．用标签识别拆下的零件，此方法将有助于随后的装配过程。

B．保存好拆下标签的零件，此方法将有助于随后的装配过程。

C．用标签标记拆下的零件，此方法将有助于子部件的安装。

（　）7．Troubleshooting information describing probable malfunctions，how to recognize those malfunctions，and the remedial action for those malfunctions shall be provided.

A．应另提供排故信息，说明可能的故障，如何识别这些故障的方法以及故障的纠正措施。

B．故障排除信息描述可能发生的故障，应另提供如何识别这些故障的方法，以及针对这些故障采取的合理措施。

C．故障排除信息描述可能发生的故障，应另提供如何识别这些故障的方法，以及针对这些故障采取的有效措施。

（　）8．在夜间操作期间，确认面板上的灯是亮的。

A．Within night operation，make sure the panel lights are dim.

B．During night operation，make sure the panel lights are bright.

C．During night operation，make sure the panel lights are dim.

（　）9．2 毫米的值是可以接受的。

A．A value of 2 mm is permitted.

B．A value of 2 mm is rejected.

C．A value of 2 mm is unacceptable.

（　）10．不要在机库里进行这个步骤。

A．DO NOT DO THIS PROCEDURE IN THE HANGAR.

B．DO NOT DO THIS PROCEDURE IN THE HANGER.

C．DO NOT DO THIS EXAMINATION IN THE HANGER.

（　）11．螺栓在法兰的对面。

A．The bolts are on the opposite side of the flange.

B．The bolts are on the rear side of the flange.

C．The bolts are on the across side of the flange.

（　）12．当使用该组件前，需先进行特定的测试程序。

A．Before you accept the unit，do the functional test procedure.

B．Before you accept the unit，do the operational test procedure.

C．Before you accept the unit，do the specified test procedure.

（　）13．When you install the part，make sure that the rounded edge is against the structure.

A．When the part is installed，make sure that the rounded edge is adjacent to the structure.

B. Make sure that the rounded edge leans opposite the structure when you install the part.

C. When you install the part, ensure that the rounded edge clings to the structure tightly.

( ) 14. The procedure excludes the removal of the piston.

A. The removal of the piston is involved in the steps.

B. The removal of the piston is not included in the procedure.

C. The procedure eliminates the piston.

( ) 15. Because of the varied mission the airplanes performed, type of care given, and age and utilization rates, ABC has determined, based on inspections, tests and analyses, that it is prerequisite to provide additional inspection requirements to further ensure that the airplanes can continue to carry the design loads it was originally certified for.

A. ABC has determined the varied mission the airplanes performed, type of care given, and age and utilization rates because providing additional inspection requirements is prerequisite to further ensure that the airplanes can continue to carry the design loads it was initially certified for based on inspections, tests, and analyses.

B. ABC has determined that providing additional inspection requirements is prerequisite to further ensure that the airplanes can continue to carry the design loads it was initially certified for based on inspections, tests, and analyses because of the varied mission the airplanes performed, type of care given, and age and utilization rates.

C. To further ensure that the airplanes can continue to carry the design loads it was initially certified for, owing to the diversified mission the airplanes performed, type of care given, and age and utilization rates, ABC has determined that providing additional inspection requirements is necessary based on inspections, tests, and analyses.

Part VI The Pillars of a Great Power

C919

On May 28, 2023, the C919 made its first commercial flight from Shanghai Hongqiao Airport to Beijing Capital Airport. This historic event is of great significance to China's aviation industry and the overall economic development, and will form a far-reaching impact.

The C919, full name COMAC C919, is a large civil jet developed by China in accordance with international civil aviation regulations and with independent intellectual property rights. It has a seat class of 158-192 and a range of 4,075-5,555 kilometers.

The domestic large aircraft C919 is safe, reliable and comfortable.

As a big aviation country, China has perfect and meticulous regulations on aircraft manufacturing, flight and maintenance. The guidance and supervision of the CAAC, the regional administration and the subordinate supervisory authority are also quite strict and efficient. All these jointly ensure the safety, order and standard of China's aviation operation.

The safety and speed of development of China's civil aviation are internationally acclaimed.

In this strict and cautious environment, the C919 project has been officially released in 2009. The prototype was rolled off the assembly line in 2016. and it made the maiden flight in 2017. Since then, it has moved to Shanghai, Xi'an, Dongying, Nanchang, Turpan, Hulun Buir and other places, and has completed a series of strict tests such as taxiing, test flight, transferring, noise, lighting, high temperature, and high cold. It has obtained a domestic Aircraft Certificate in 2022 and has been officially put into operation.

The C919 has the highest design safety standards in its class. At the personal level of passengers, there is no gap between C919 and Airbus A320 and Boeing 737, which are the top products of the same kind, and it's completely safe to take while traveling.

Of course, in comparison, Boeing 737 has produced more than 10,000 aircraft for more than 50 years, and Airbus A320 has produced more than 8,000 aircraft for more than 30 years. As a new recruit, C919 still has a long way to go. However, with China's strength and determination, as well as the efforts and persistence of one by one professionals, I believe that the C919 can also be like Huawei smart phones, not only can survive in the cracks, but also will slowly grow into an important and even top member of the world one day.

Keys for Reference

译文

Module

02 Regulations and Organizations

Project 1 Service Documents

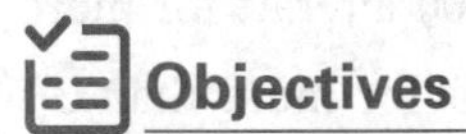

Objectives

To know the main maintenance documents.

To understand the airworthiness limitation precautions.

To master the usage of turning tools.

To perceive the importance of human factors in aviation.

Part I Aviation Situational Communication

Aircraft Communication System

1. Read aloud the topic-related sentences for aviation communication and try to match them with their Chinese meaning.

雷达管制通话常见表达

_______ (1) Maintaining FL310. (FL=Flight Level)

_______ (2) Request further climb.

_______ （3）Maintain present speed.

_______ （4）Continue descent to 3，000 feet，QNH 1012.

_______ （5）Right heading 330，descending to 3，000 feet，cleared for ILS approach Runway 36R.

Listen

_______ （6）Maintaining FL350，cleared to destination，flight planned route.

_______ （7）Delay not determined due to runway obstruction.

_______ （8）No delay expected.

A．请求进一步上升。
B．继续下降到 3 000 英尺，修正海压 1 012。
C．保持高度 350，可以飞往目的地机场，按照计划的飞行航路。
D．保持当前速度。
E．右转航向 330，下降到 3 000 英尺，可以盲降进近，跑道 36 右。
F．由于跑道障碍物，延误待定。
G．保持高度 310。
H．预计无延误。

注：QNH=Query Normal Height，1 012 读作 one zero one two。

修正海压高度读法：在 QNH 后 + 高度数值。

例：修正海压 1 200 米。

读法：one thousand two hundred meters on QNH.

2．Listen to some sentences for aviation communication carefully and try to fill in the blanks.

（1）_______ not determined due to runway obstruction.

（2）Ready for _______.

（3）_______ to FL280.

（4）Request further _______.

（5）_______ present speed.

Listen

Part II Aviation Reading

2.1 Aviation Reading

Pre-reading Questions

1．Why do aircrafts require various types of maintenance?

2. What are the maintenance documents?

---

3. What do the maintenance documents include?

Maintenance Documents

For the civil aviation industry, safety is the top priority. Before being put into use, the civil aircraft must have initial airworthiness. However, the aircraft system' s performance may deteriorate or even be directly damaged during use, resulting in malfunctions that cannot meet continuing airworthiness requirements. Therefore, ensuring an aircraft fly safely requires various types of maintenance (preventive, snags, scheduled, unscheduled, etc.) to be performed on the aircraft at various periodicities based on days, flying hours, aircraft engine starts, aircraft landings, aircraft cycles, etc.

The basic quality of aircraft during the entire operation and use process must be maintained and restored by the user and maintenance organization in accordance with various approved usage specifications, maintenance rules and standards, which collectively referred to as continuing maintenance documents. The continuing airworthiness maintenance documents include technical materials approved by aviation authorities (e.g., CAAC, FAA), technical materials provided by aircraft manufacturers, maintenance documents from the aviation operator, and maintenance manuals from the maintenance organization.

Technical materials approved by aviation authorities

The continuing airworthiness technical materials approved by aviation authorities (e.g., CAAC) include Airworthiness Limitation Items (ALI), Maintenance Review Board Report (MRBR), Certification Maintenance Requirements (CMR), Master Minimum Equipment List (MMEL), Structural Repair Manuals (SRM), and important repair and modification plans. The aviation authorities issued and provided maintenance related documents, including aviation regulations (FAR or CCAR), Advisory Circular (AC), Airworthiness Directive (AD), and Notices of Proposed Rulemaking (NPRM), etc.

The following is an introduction to common aviation regulations.

CCAR

CCAR stands for China Civil Aviation Regulations, which are formulated and issued by the civil aviation authority of China—Civil Aviation Administration of China (CAAC) in accordance with the Civil Aviation Law of the People' s Republic of China and the Convention on International Civil Aviation (also known as Chicago Convention). The regulations cover various aspects of civil aviation, including aircraft management, personnel licenses participating in civil aviation activities, airport management, navigation management, aviation operations, air traffic management, etc. Any individual or organization engaged in civil aviation activities within the territory of China must comply with the various regulations of CCAR.

FAR

FAR stands for Federal Aviation Regulations, which are prescribed by the Federal Aviation Administration (FAA) governing all aviation activities in the United States. The FARs are part of title 14 of the Code of Federal Regulations (CFR).

Technical materials provided by aircraft manufacturers

The continuing airworthiness technical documents provided by the manufacturers (manufacturer's service documents) mainly include Instructions for Continued Airworthiness (ICA), engine/propeller manufacturer Maintenance Manuals / Overhaul Manuals (MM/OM), Component Maintenance Manuals (CMM), Service Bulletins / Service Letters (SB/SL), etc. Among them, ICA contains Aircraft Maintenance Manual (AMM), Schematic Diagram Manual (SDM), Wiring Diagram Manual (WDM), Fault Isolation Manual (FIM), Fault Reporting Manual (FRM), Illustrated Parts Catalogue (IPC), Structural Repair Manual (SRM), Master Minimum Equipment List (MMEL), Dispatch Deviation Guide (DDG), Configuration Deviation List (CDL), Maintenance Plan Data (MPD), Vendor Manual (VM), Task Card (TC), etc.

The following is an introduction to common technical documents.

AMM

Aircraft Maintenance Manual (AMM, Fig. 2-1) is the formal document which details the way in which all maintenance tasks carried out on an aircraft shall be accomplished. This includes items such as lubrication system functional checks and servicing of the airplane but usually excludes structural repairs and modifications.

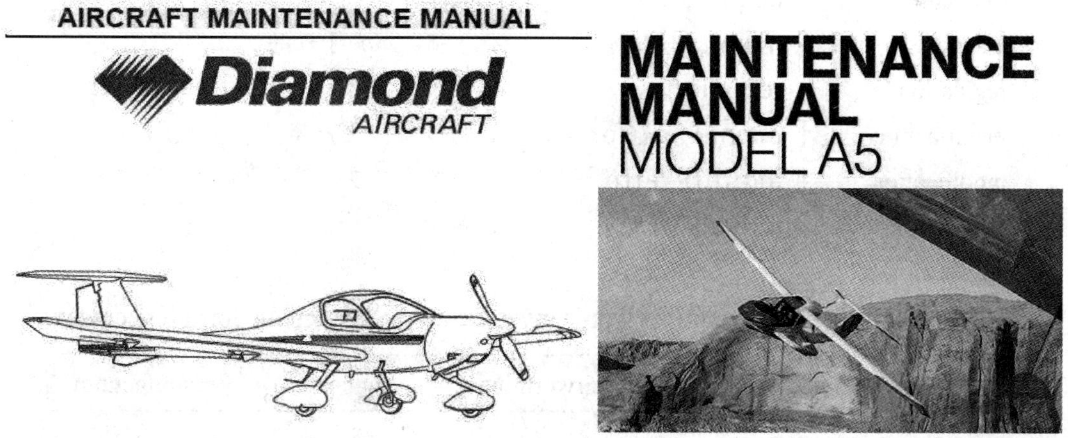

Fig. 2-1 AMM

CMM

A Component Maintenance Manual (CMM) is a formal document which details the way in which off-aircraft maintenance tasks on the specified component shall be accomplished. The maintenance tasks contained in these manuals include procedures for restoring a structural component to a serviceable state and re-working and refinishing procedures are often provided in any appropriate CMM.

SB

A Service Bulletin（SB）is the document used by manufacturers of aircraft, their engines or their components to communicate details of modifications which can be embodied in aircraft.

Maintenance documents from the aviation operator

(The maintenance & engineering management manual of aviation operators is the basic document for how aviation operators assume the airworthiness responsibility of aircraft and implement maintenance management.) It is the overall outline and writing basis of aviation operator maintenance management documents. The chapters directly related to maintenance work mainly include three parts: maintenance plan (independently compiled), technical materials (task card), and execution of maintenance and modification.

Maintenance manuals from the maintenance organization

The maintenance manual from the maintenance organization should meet both the requirements of airworthiness regulations and the requirements of aviation operators.

Words and Expressions

airworthiness	[ˈerwɜːrðinəs]	*n.* 适航性
authority	[əˈθɔːrəti]	*n.* 当局，权威
specification	[ˌspesɪfɪˈkeɪʃn]	*n.* 规范，说明书
approve	[əˈpruːv]	*v.* 批准，同意
overhaul	[ˈoʊvərhɔːl]	*n.* 大修
manual	[ˌmænjuˈ]	*n.* 手册 *adj.* 手动的
regulation	[ˌregjuˈleɪʃn]	*n.* 法规，章程
manufacturer	[ˌmænjuˈfæktʃərər]	*n.* 制造商；生产商
modification	[ˌmɑːdɪfɪˈkeɪʃn]	*n.* 修改，改变
lubrication	[ˌluːbrɪˈkeɪʃn]	*n.* 润滑

Fill in the blanks with the proper words given below, changing the form if necessary.

manual	regulation	airworthiness	approve	manufacturer

1. China's homegrown C919 had received its ______ certificate and was delivered to China Eastern Airlines in 2022.
2. The course is ______ by the Department for Education.
3. He has read the ______ carefully so that the product won' t be misapplied.
4. The company was fined £ 20, 000 for breaching safety ______.
5. It is important to follow the _____ instructions.

2.2 Manual Reading

中国民用航空局  CIVIL AVIATION ADMINISTRATION OF CHINA

CAAC

适 航 指 令

AIRWORTHINESS DIRECTIVE

本指令根据中国民用航空规章《民用航空器适航指令规定》（CCAR-39）颁发，内容涉及飞行安全，是强制性措施。如不按规定完成，有关航空器将不再适航。

There are some tasks for you to fulfill. You should read the Manual Reading materials carefully and then finish the tasks.

Airworthiness Limitation Precautions

General

(1) Critical Design Configuration Control Limitations (CDCCLs)

① All occurrences of CDCCLs found in this chapter of the AMM are identified by this note after each applicable CDCCL design feature:

NOTE: CDCCL—Refer to the task: Airworthiness Limitation Precautions, TASK 05-00-00-910-801, for important information on Critical Design Configuration Control Limitations (CDCCLs).

② Design features that are CDCCLs are defined and controlled by Special Federal Aviation Regulation (SFAR) 88, and can be found in Section 9 of the Maintenance Planning Data (MPD) document. CDCCLs are a means of identifying certain design configuration features intended to preclude a fuel tank ignition source for the operational life of the airplane.

CDCCLs are mandatory and cannot be changed or deleted without the approval of the FAA office that is responsible for the airplane model type certificate, or applicable regulatory agency. A critical fuel tank ignition source prevention feature may exist in the fuel system and its related installation or in systems that, if a failure condition were to develop, could interact with the fuel system in such a way that an unsafe condition would develop without this limitation. Strict adherence to configuration, methods, techniques, and practices as prescribed is required to ensure the CDCCL is complied with. Any use of parts, methods, techniques or practices not contained in the applicable CDCCL must be approved by the FAA office that is responsible for the airplane model type certificate, or applicable regulatory agency.

(2) Airworthiness Limitation Instructions (ALIs)

① All occurrences of fuel tank system ALIs found in this chapter of the AMM are identified by this step after the General section in the applicable ALI inspection task:

ALI - Refer to the task: Airworthiness Limitation Precautions, TASK 05-00-00-910-801, for important information on airworthiness limitation instructions (ALIs).

② Inspection tasks that are ALIs are defined and controlled by Special Federal Aviation

Regulation（SFAR）88，and can be found in section 9 of the Maintenance Planning Data（MPD）document. These ALIs identify inspection tasks related to fuel tank ignition source prevention which must be done to maintain the design level of safety for the operational life of the airplane. These ALIs are mandatory and cannot be changed or deleted without the approval of the FAA office that is responsible for the airplane model type certificate，or applicable regulatory agency. Strict adherence to methods，techniques and practices as prescribed is required to ensure the ALI is complied with. Any use of methods，techniques or practices not contained in these ALIs must be approved by the FAA office that is responsible for the airplane model type certificate，or applicable regulatory agency.

Critical Design Configuration Control Limitations（CDCCLs）

WARNING: OBEY THE MANUFACTURER'S PROCEDURES WHEN YOU DO MAINTENANCE THAT HAS AN EFFECT ON A CDCCL. IF YOU DO NOT OBEY THE PROCEDURES，IT CAN INCREASE THE RISK OF A SOURCE OF FUEL TANK IGNITION. INJURIES TO PERSONNEL，AND DAMAGE TO EQUIPMENT CAN OCCUR IF THERE IS A FIRE OR EXPLOSION.

Make sure you follow the procedures for items identified as CDCCLs.

Airworthiness Limitation Instructions（ALIs）

WARNING: OBEY THE MANUFACTURER'S PROCEDURES WHEN YOU DO ANY MAINTENANCE THAT MAY AFFECT AN ALI. IF YOU DO NOT FOLLOW THE PROCEDURES，IT CAN INCREASE THE RISK OF A FUEL TANK IGNITION SOURCE.

Make sure you follow the procedures for items identified as ALIs.

Selected from AMM

Words for Reference

airworthiness	['erwɜːrðinəs]	*n.* 适航性
precaution	[prɪ'kɔːʃn]	*n.* 注意事项
configuration	[kənˌfɪgjə'reɪʃn]	*n.* 结构，构造
preclude	[prɪ'kluːd]	*v.* 阻止，妨碍
mandatory	['mændətɔːri]	*adj.* 强制性的；法定的
approval	[ə'pruːvl]	*n.* 批准；同意
installation	[ˌɪnstə'leɪʃn]	*n.* 装置，安装
ignition	[ɪg'nɪʃn]	*n.* 点火，点火装置
fuel tank		油箱
regulatory agency		监管机构
CDCCLs（Critical Design Configuration Control Limitations）		重要设计构型控制限制
ALIs（Airworthiness Limitation Instructions）		适航性限制说明
SFAR（Special Federal Aviation Regulation）		特别航空管理规定
MPD（Maintenance Planning Data）		维修计划数据

Reading Tasks

1. Judge the following statements according to the contents of the AMM.

(1) When you do any maintenance that may affect an ALI, it's not necessary to obey the manufacturer's procedures. ( )

(2) Only CDCCLs are mandatory and cannot be changed or deleted without the approval of the FAA office or applicable regulatory agency. ( )

(3) Inspection tasks that are ALIs and design features that are CDCCLs are defined and controlled by Special Federal Aviation Regulation (SFAR) 88. ( )

2. Fill in the blanks according to the above reading materials.

(1) Any use of methods, techniques or practices not contained in these ALIs must be ______________ by the FAA office that is responsible for the airplane model type certificate.

(2) A critical ______________ source prevention feature may exist in the fuel system and its related installation or in systems.

(3) These ALIs are ______________ and cannot be changed without the approval of the FAA office or applicable regulatory agency.

Translate the following sentences into Chinese.

1. Make sure you follow the procedures for items identified as CDCCLs.

____

2. Obey the manufacturer's procedures when you do any maintenance that may affect an ALI.

____

3. Injuries to personnel, and damage to equipment can occur if there is a fire or explosion.

____

4. If you do not follow the procedures, it can increase the risk of a fuel tank ignition source.

____

5. Any use of methods, techniques or practices not contained in these ALIs must be approved by the FAA office that is responsible for the airplane model type certificate, or applicable regulatory agency.

____

Translate the following abbreviations into corresponding Chinese.

1. SFAR____
2. ALIs____
3. CDCCLs____
4. MPD____

Part III Tools and Equipments

Turning Tools

1. Match the words or phrases with the translations.

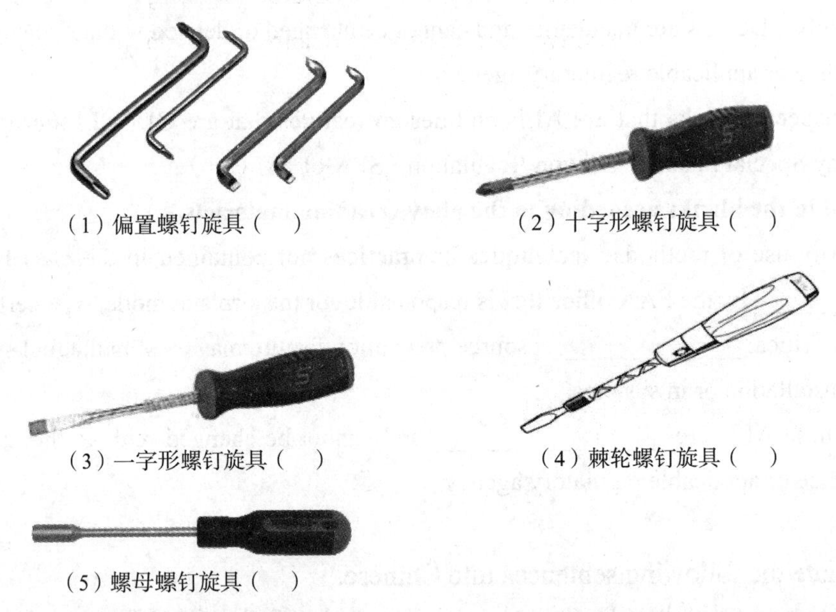

（1）偏置螺钉旋具（　）

（2）十字形螺钉旋具（　）

（3）一字形螺钉旋具（　）

（4）棘轮螺钉旋具（　）

（5）螺母螺钉旋具（　）

A. Ratcheting screwdriver
B. Flathead or slotted screwdriver
C. Offset screwdriver
D. Phillips screwdriver
E. Nut screwdriver

2. Fill in the words or phrases in the blanks.

Nut screwdriver
Offset screwdriver
Phillips screwdriver
Ratcheting screwdriver
Flathead or slotted screwdriver

（1）It is a tool with a wedge-shaped flat tip that is designed to fit into the notch on a slotted screw. When choosing a screwdriver for a slotted screw, the blade should fill it at least 75% of the slot.

（2）It is the most common cross point screwdriver, which has an X-shaped tip that is made to fit inside the X-shaped notch of a compatible Phillips head screw. The shape of the screwdriver head provides better control and driving power when you are driving or removing screws.

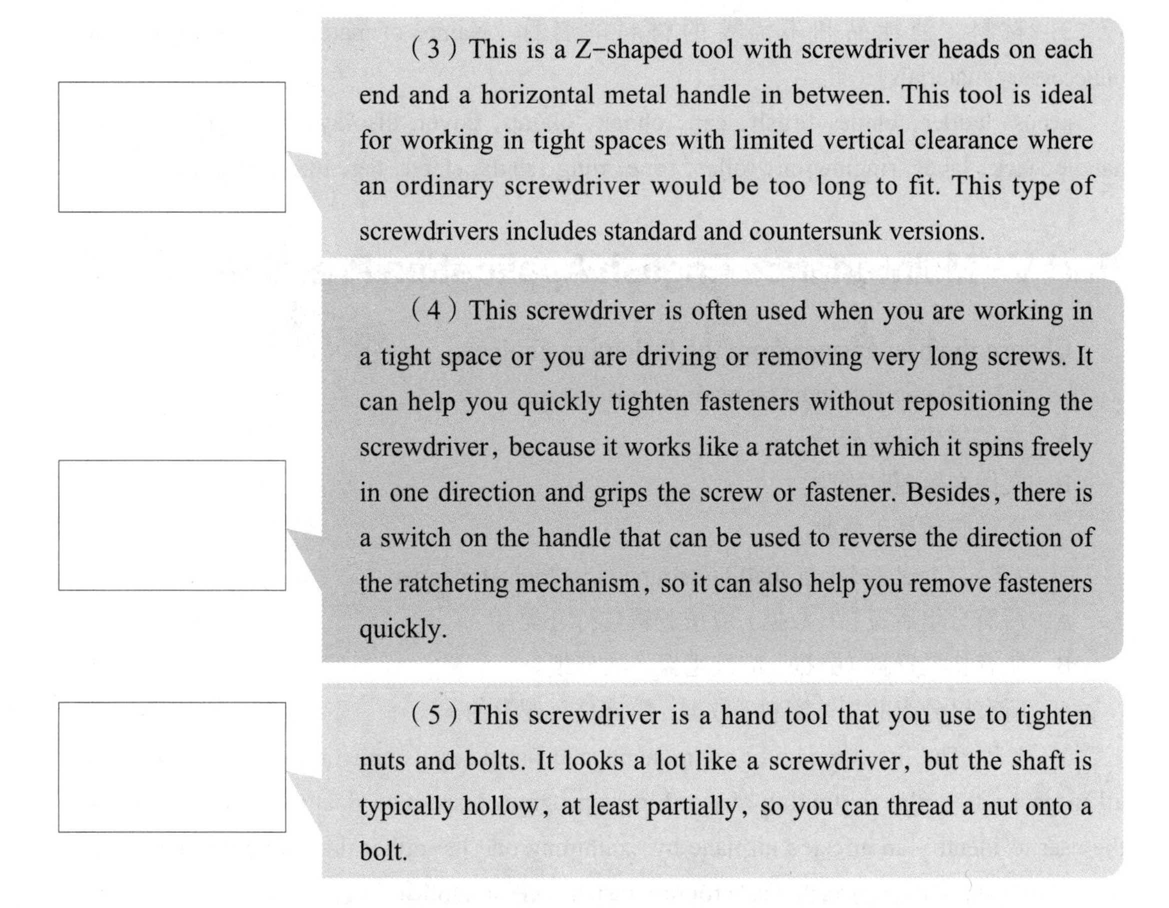

(3) This is a Z-shaped tool with screwdriver heads on each end and a horizontal metal handle in between. This tool is ideal for working in tight spaces with limited vertical clearance where an ordinary screwdriver would be too long to fit. This type of screwdrivers includes standard and countersunk versions.

(4) This screwdriver is often used when you are working in a tight space or you are driving or removing very long screws. It can help you quickly tighten fasteners without repositioning the screwdriver, because it works like a ratchet in which it spins freely in one direction and grips the screw or fastener. Besides, there is a switch on the handle that can be used to reverse the direction of the ratcheting mechanism, so it can also help you remove fasteners quickly.

(5) This screwdriver is a hand tool that you use to tighten nuts and bolts. It looks a lot like a screwdriver, but the shaft is typically hollow, at least partially, so you can thread a nut onto a bolt.

Part IV Simplified Technical English

Technical Names

技术名称

★规则：可以使用包含在技术名称类别中的单词。

技术名称是与特定类别相关的词。字典中没有将技术名称作为认可的词，因为技术名称太多，而且每个制造商使用不同的技术名称。STE 提供了一个类别列表和示例，分为 19 个类别，以下以其中三个类别为例。

1. 官方零件信息中的名称（Names in the official parts information）（for example, Illustrated Parts Catalog or engineering drawing）:

bolt, cable, clip, conductor, contact, engine, ferry tank, filter, hatch, light, logo, oil seal, pipe, propeller, screw.

2. 工具和支持设备的名称、部件和位置（Names, components and locations of vehicles and machines）:

aircraft, aircraft carrier, airframe, airplane, bicycle, cabin, car, cargo compartment, cargo hold, cockpit, deck, engine room, fuselage, helicopter, galley, lifeboat, overhead panel.

3．材料、消耗品和不需要的材料的名称（Names of materials, consumables, and unnecessary materials）:

access ladder，blade，brush，cap，chock，clamp，cover, display，file，gauge（gage），handle，jack，label，rigging pin，roller，rope，rung，shaft，stand，tag，torque wrench.

Part V　Maintenance English Application Practice

Choose the best answer from the following choices.

（　）1．Documents must be readily revisable.

A．文件需要随时修改。

B．文件必须随时修改。

C．文件必须易于修改。

（　）2．All related data shall be grouped in a logical manner.

A．所有分组后资料（数据）应按逻辑进行分类。

B．所有相关资料（数据）应按逻辑方式分组。

C．所有已分组的相关资料（数据）应该是有逻辑的。

（　）3．The listed model or components and listed serials/units are to be cumulative for all inspections in the continuing airworthiness program inspection document. This will allow the user to identify an affected airplane by examining one list rather than having to check each individual continuing airworthiness program inspection for applicability.

A．在持续适航方案检查文件中，列出的型号或部件和列出的系列号 / 组件的所有检查应采取分列方式。这将允许用户通过查看一个清单来识别受影响的飞机，不需要去查看每个单独持续适航方案检查的合理性。

B．在持续适航方案检查文件中，列出的型号或部件和列出的系列号 / 组件的所有检查应采取累计方式。这将允许用户通过查看一个清单来识别受影响的飞机，而不是必须去查看每个单独持续适航方案检查的适用性。

C．在持续适航方案检查文件中，列出的型号或部件和列出的系列号 / 组件的所有检查应采取计数方式。这将允许用户通过查看一个清单来识别受影响的飞机，以及去查看每个单独持续适航方案检查的有效性。

（　）4．The Continuing Airworthiness Program（CAP）inspection document is inspection data that，when combined with the operator's existing maintenance program，will help maintain the structural integrity and continued airworthiness of the ABC ××× series airplanes.

A．如果与营运人现有的维修方案相结合，持续适航方案检查文档（CAP）作为一种检查资料，将有助于保持 ABC ××× 系列飞机的结构稳定性和持续适航性。

B．持续适航方案检查文档（CAP）是一种检查资料，与营运人现有的维修方案相结合，将有助于保持 ABC ××× 系列飞机的结构完整性和持续适航性。

C．持续适航方案检查文档（CAP）是一种与营运人现有的维修方案相结合的检查资料，将有助于保持 ABC ××× 系列飞机的结构辨识度和持续适航性。

(   ) 5．将通过统一解释根据本规范编制的持续适航方案中包含的检查要求来提高安全性。

A．Standardization will through uniform interpretation of inspection requirements in accord with this specification contained in continuing airworthiness programs prepared to enhance safety.

B．Standardization will enhance safety through uniform interpretation of inspection requirements contained in continuing airworthiness programs prepared in accord with this Specification.

C．Standardization will in accord with this specification through uniform interpretation of inspection requirements contained in continuing airworthiness programs prepared to enhance safety.

(   ) 6．不包括应包含在使用结构章节中机身的整体结构。

A．Include integral structure with the airframe, which shall be included in the applicable structure chapter.

B．Does not include integral structure with the airframe, which shall be included in the applicable structure chapter.

C．Integral structure with the airframe, which shall be included in the applicable structure chapter, does not include.

(   ) 7．巡航性能介绍的格式由飞机制造厂商自行决定。

A．The form of the climb performance presentation is decided by the airplane manufacturer.

B．The format of the cruise performance presentation is at the discretion of the airplane manufacturer.

C．The form of the cruise performance presentation is at the discretion of the airplane.

(   ) 8．本文件向适用的飞机维修手册提供补充信息。

A．This document provides optional information to the befitting airplane maintenance manual(s).

B．This document provides supplemental information to the applicable airplane maintenance manual(s).

C．This document provides extra information to the accessible airplane maintenance manual(s).

(   ) 9．提供给客户的所有技术数据中的术语应保持一致。

A．Nomenclature shall be similar throughout all technical data obtained by the customer.

B．Nomenclature shall be consistent throughout all technical data given to the customer.

C．Nomenclature must be similar throughout all technical data provided to the customer.

(   ) 10．服务通告中程序的执行是强制的。

A．It is standard to do the procedure in this service bulletin.

B. It is necessary to do the procedure in this service bulletin.

C. It is mandatory to do the procedure in this service bulletin.

( ) 11. In 1944, the British Civil Airworthiness Requirements was developed to form the code of the airworthiness standards through which the requirements of the Air Navigation Orders and Regulations would be met.

A. 1944 年，随着空中航行法令和规章的各种要求得以实现，英国民用适航机构形成了标准的法典。

B. 1944 年，英国制定了民用适航要求来形成空中航行指令和规则守则。

C. 1944 年，随着英国民用适航要求的制定，形成了适航标准的法典，通过该法典，空中航行法令和规章的各种要求得以实现。

( ) 12. The purpose of such terms is to provide definitions which are of practical value for technical communication.

A. The intention of such terms is to provide definitions which have practical value for technical communication.

B. The propose of such terms is to provide definitions with practical value for technical communication.

C. The aim of such terms is to provide definitions for practicing valuable technical communication.

( ) 13. If you are not sure if the unit is serviceable, do a standard serviceability test.

A. If you have no doubt about the serviceability of the unit, perform a standard serviceability test.

B. If you have no knowledge about the serviceability of the unit, do a standard serviceability test.

C. If there are doubts about the serviceability of the unit, carry out a standard serviceability test.

( ) 14. The rules contained in this specification shall serve as guides for format, style and method of presentation for all material included in maintenance manuals prepared by manufacturers.

A. In terms of format, style and method of presentation of the rules written in this specification, the material contained in maintenance manuals designed by manufacturers should be prepared under the guidance of it.

B. In terms of format, style and method of presentation, the material contained in maintenance manuals designed by manufacturers should be made in accordance with the rules written in this specification.

C. The material contained in maintenance manuals designed by manufacturers should be prepared under the guidance of the rules written in this specification in terms of format, style and method of presentation.

(　) 15. The listing of CAP inspections contains all continuing all worthiness program Inspections in the CAP, in numerical order by CAP inspection number, with full title, date, and effectivity and inspection compliance.

A. In the Listing of CAP Inspections, apart from full title, date, effectivity and inspection compliance, all continuing airworthiness program inspections in the CAP are excluded in, numerical order on the basis of CAP inspection number.

B. Along with full title, date, effectivity and inspection compliance, all continuing airworthiness program inspections in the CAP are excluded in numerical order in the light of CAP inspection number, in the listing of CAP Inspections.

C. All continuing airworthiness program inspections in the CAP are covered, in numerical order by CAP inspection number, in the listing of CAP inspections along with full title, date, and effectivity and inspection compliance.

Part VI Safety and Regulations

The Importance of Human Factors in Aviation

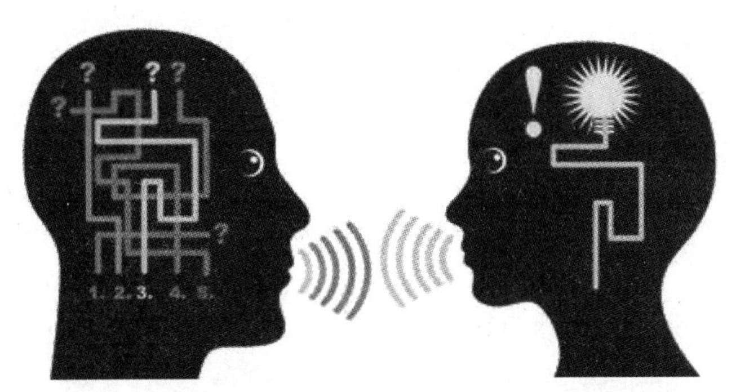

Xi Jinping, General Secretary of the Communist Party of China (CPC) Central Committee, stressed on the 20th CPC National Congress to accelerate the efforts to modernize the country's national security system and capacity, so as to safeguard the new development pattern with a new security architecture. In aviation industry, keep passengers and flight crew safe while flying, safety always comes first.

Aviation is a complex business and involves the participation of people in more spheres than one. From the manufacturer, maintenance, ground support, ATC, inflight to the flight crew, and even passengers, every agency plays a role from the safe takeoff to landing of every flight. As human beings, due to the limitations of their psychology, physical strength, intelligence, reaction ability, etc., they are likely to make mistakes when they have a heavy workload or struggle with fatigue. According to Boeing, human factors are involved in more than 70 percent of aircraft accidents and affect all areas of aviation operations, including flight, maintenance, and air traffic management. This is why the understanding of human factors is extremely important. By understanding the properties of human capability and ensuring successful application of human factor principles into the aviation working environment, human errors will be reduced.

Your opinion

In your opinion, what are the main human factors that affect the maintenance quality in the process of aviation maintenance? And how to effectively improve maintenance skills?

Project 2 International Aviation Organizations

Objectives

To know the main international civil aviation organization.

To understand the dimensions and areas in maintenance practices.

To master the usage of pounding tools.

To perceive the craftsman spirit of Chinese aviation.

Part I Aviation Situational Communication

1. Read aloud the topic-related sentences for aviation communication and try to match them with their Chinese meaning.

Common expressions for approach control calls

_______ （1）Request visual approach.

_______ （2）Continue approach Runway 36R，maintain visual separation with preceding traffic.

_______ （3）Roger，request continue approach.

_______ （4）Established on the localizer.

_______ （5）Contact tower 116.8.

_______ （6）Request holding instructions.

_______ （7）Passing outer marker.

_______ （8）Runway in sight.

Listen

A．明白，请求继续进近。

B．请求等待指令。

C．联系塔台 116.8。

D．请求目视进近。

E．继续进近，跑道 36 右，与前机保持目视间隔。

F．过外指点标了。

G．看见跑道了。

H．已经建立好航向道。

2. **Listen to some sentences for aviation communication carefully and try to fill in the blanks.**

(1) Airfield in sight, request __________approach.

(2) Request holding _________.

(3) Number one, __________ Tower 116.8.

(4) Fully established __________ 20.

(5) Continue __________, prepare for possible go-around.

Listen

Part II Aviation Reading

2.1 Aviation Reading

Pre-reading Questions

1. How many agencies does the passage mention? And what are they?

____

2. What is "ICAO"?

____

3. When is the European Aviation Safety Agency (EASA) established and what is its goal?

____

International Civil Aviation Organization

The International Civil Aviation Organization (ICAO) is a specialized agency of the United Nations. It was established in 1947 to promote the safe and orderly development of civil aviation around the world. Headquartered in Montreal, Canada, ICAO sets international air transport standards and regulations and serves as a vehicle for cooperation in the field of civil aviation among its 193 contracting States (up to 2022). The purpose of ICAO is to develop the principles and techniques of international navigation and to promote the planning and development of international air transport. For example, to encourage the design and operation of aircraft for peaceful purposes, and to encourage the development of air routes, airports and navigation facilities for international civil aviation applications.

International Air Transport Association

The International Air Transport Association (IATA) is a trade association of the world's airlines founded in 1945. It is a large international organization composed of airlines from all over the world. It is an unofficial organization.

IATA's mission is to promote safe, normal and economical air transport for the benefit of the people of the world. IATA supports airline activity and helps formulate industry policy and standards. In addition to setting technical standards for airlines, IATA manages issues arising in air transport such as fares and the transport of dangerous goods. It coordinates and communicates policies between governments through air transport enterprises and solves practical operational problems.

The basic functions of the association include: harmonization of international air transport regulations, business agent, financial settlement between air carriers, technical cooperation, participation in airport activities, coordination of international air passenger and cargo rates, aviation legal work, and assistance in training senior and specialized personnel for airlines in many areas.

The Federal Aviation Administration

The Federal Aviation Administration (FAA) is the largest modern transportation agency and a governmental agency of the United States. Created in August 1958, the FAA replaced the former Civil Aeronautics Administration (CAA) and later became an agency within the U.S. Department of Transportation.

Its main responsibilities include: Promoting civil aviation safety management; to encourage and develop civil aviation, including new aviation technologies; developing and operating air traffic control and navigation systems for civil and military aircraft; research and development systems and civil aviation airspace; to develop and implement controls on aircraft noise and other programme environmental impacts on civil aviation; American commercial space transportation management, etc.

European Aviation Safety Agency

The European Aviation Safety Agency (EASA) is an agency of the European Union (EU) with responsibility for civil aviation safety. It was established in 2003 with the goals of maximizing the safety of citizens and promoting the development of aviation in the European Union.

The main responsibilities of the EASA agency are to draft civil aviation safety regulations, provide technical support to the European Union, and provide technical assistance to the relevant international agreements.

In addition, the agency carries out operational certification related to aviation safety, such as certification of aviation products and organizations related to design, manufacture and maintenance. Establish general rules for the continuous protection of civil aviation safety and environmental protection, establish continuous airworthiness standards for all aircraft types, and set safety standards for institutions and personnel responsible for the design, manufacture and maintenance of aircraft.

Words and Expressions

agency	[ˈeɪdʒənsi]	*n.* 服务/代理/经销机构；(政府)专门机构
standard	[ˈstændəd]	*n.* (品质的)标准，水平，规范
regulation	[ˌreɡjuˈleɪʃn]	*n.* 规章制度，规则；(运用规则条例的)管理，控制
navigation	[ˌnævɪˈɡeɪʃn]	*n.* 导航；航行，航海；航运，水上运输
cooperation	[kəʊˌɒpəˈreɪʃn]	*n.* 合作，协作；协助，配合
coordination	[kəʊˌɔːdɪˈneɪʃn]	*n.* 协调，配合；身体的协调性
cargo	[ˈkɑːɡəʊ]	*n.* (船或飞机装载的)货物
military	[ˈmɪlətri]	*adj.* 军事的，军队的；*n.* 军人，军方
implement	[ˈɪmplɪment]	*v.* 执行，贯彻；*n.* 工具，器具
certification	[ˌsɜːtɪfɪˈkeɪʃn]	*n.* 证明，资质证书
manufacture	[ˌmænjuˈfæktʃə(r)]	*v.* (用机器大量)生产，制造；*n.* 工业品；制造业
ICAO		*abbr.* 国际民用航空组织
IATA		*abbr.* 国际航空运输协会

FAA　　　　abbr. 美国联邦航空管理局
EASA　　　　abbr. 欧洲航空安全局

Fill in the blanks with the proper words given below, changing the form if necessary.

agency	navigation	cooperation	coordination	implement

(1) The _______ is a research and development arm of the department of defense.
(2) He emphasized the wider issue of superpower ______.
(3) Team management provides for this ______.
(4) Leadership is about the ability to ______ change.
(5) We don't have that problem, because we have ______ systems like GPS.

2.2 Manual Reading

There are some tasks for you to fulfill. You should read the Manual Reading materials carefully and then finish the tasks.

DIMENSIONS AND AREAS—MAINTENANCE PRACTICES

General

This chapter gives the principal dimensions for the wing, ailerons, flaps, horizontal stabilizer surfaces, vertical stabilizer surfaces, and body. It also gives areas for the wing and tail surfaces. There are many station lines, such as wing, vertical tail surfaces, engine nacelle and access doors and panels.

Reference Planes and Lines

The airplane is divided into stations, waterlines, and buttock lines. They are measured in inches. They will help you quickly identify the location of components, the center of gravity and the weight distribution.

Standard Abbreviations and Definitions

B STA, BS, or STA Body (Fuselage) Station.

A plane that is perpendicular to the fuselage centerline. It is measured from a point 130, 00 inches forward of the nose.

BBL or BL Body (Fuselage) Buttock Line

A vertical plane that is parallel to the vertical centerline plane, BBL 0.00. It is found by its perpendicular distance from the fuselage centerline plane. (It is a measurement of width.)

LBL Left Buttock Line

RBL Right Buttock Line

BWL or WL Body (Fuselage) Waterline

A plane that is perpendicular to the BBL plane, parallel to the fuselage centerline. It is measured from a parallel imaginary plane，BWL 0.00，148, 5 inches below the lowest fuselage surface.

BRP Body（Fuselage）Reference Plane

A plane that is perpendicular to the BBL plane and goes through BWL 208.10，the top of the main deck floor beams.

Selected from AMM

Words for Reference

dimension	[daɪ'menʃn]	*n.* 尺寸
nacelle	[næ'sel]	*n.*（飞机的）发动机舱
aileron	['eɪləˌrɒn]	*n.* 副翼
station line		站位线
buttock line		纵剖线
water line		水线
gravity	['grævɪtɪ]	*n.* 重力
abbreviation	[əˌbrivi'eɪʃn]	*n.* 缩写
stabilizer	['steɪbɪlaɪzə]	*n.*（航空器的）稳定器
perpendicular	[ˌpɜːpən'dɪkjʊlə]	*adj.* 垂直的

Reading Tasks

1. Judge the following statements according to the contents of the AMM.

（1）The stations，waterlines，and buttock lines are measured in centimeters.（ ）

（2）Body（Fuselage）station is parallel to the fuselage centerline.（ ）

（3）BWL 0.00 is an imaginary plane，which is 148, 5 inches below the lowest fuselage surface.（ ）

2. Fill in the blanks according to the above reading materials.

（1）The reference plane and lines will help you quickly identify the __________ of components，the center of gravity and the weight distribution.

（2）Fuselage Buttock Line is a ______plane，BBL 0.00 is found by its perpendicular distance from the fuselage centerline plane.

（3）_________________ is perpendicular to the BBL plane and goes through BWL 208.10，the top of the main deck floor beams.

Translate the following sentences into Chinese.

1. This chapter gives the principal dimensions for the wing，ailerons，flaps，horizontal stabilizer surfaces，vertical stabilizer surfaces，and body.

____

2. The airplane is divided into stations, waterlines, and buttock lines.

____

3. They are measured in inches.

____

4. A plane that is perpendicular to the fuselage centerline.

____

5. It also gives areas for the wing and tail surfaces.

____

Translate the following abbreviations into corresponding Chinese.

1. STA______________________
2. BBL______________________
3. LBL______________________
4. BRP______________________
5. WL______________________

Part III　Tools and Equipments

Pounding Tools

1. Match the words or phrases with the translations.

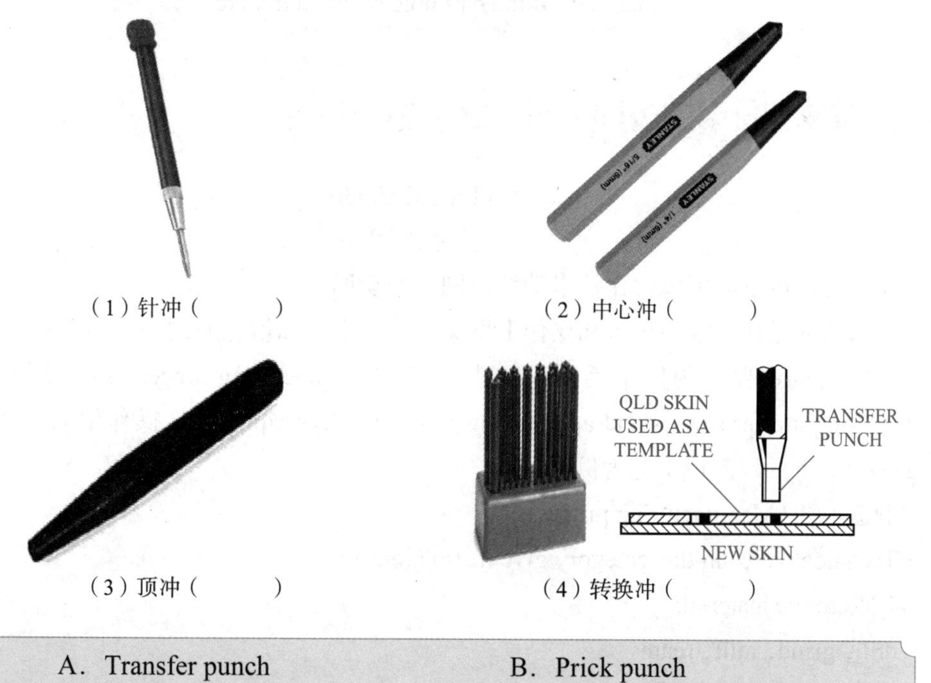

(1) 针冲 (　　　)　　(2) 中心冲 (　　　)

(3) 顶冲 (　　　)　　(4) 转换冲 (　　　)

A. Transfer punch　　B. Prick punch

C. Center punch　　D. Pin punch

2. Fill in the words or phrases in the blanks.

Transfer punch	Prick punch
Center punch	Pin punch

(1) The top of punch is flat, and the punching rod is transitional Conical.The deformation is small when the force is applied.

(2) It is used to mark dimension and locations on sheet metal. The tip of the punch is relatively thin, and do not use the hammer to knock heavily, so as not to damage the material and punch tip.

(3) It is used to make an indentation in a sheet metal as an alignment mark for starting a drill. The purpose is to position the drill hole and prevent the bit from running out of position when drilling.

(4) The position of the hole on the template or plate is accurately converted to the material to be processed. Its characteristic is that the diameter of the punch rod is equal to the diameter of the nail hole on the template.

Part IV Simplified Technical English

Technical Verbs
技术动词

★规则：可以使用包含在技术动词类别中的动词。

技术动词是在特定的技术和操作上下文中给出指示和信息的词。

技术动词主要分为四个类别：制造工艺（Manufacturing processes），计算机处理和应用（Computer processes and applications），描述（Descriptions），操作语言（Operational language）。下面以制造工艺为例。

制造工艺（Manufacturing processes）

Technical verbs in this category give instructions to

1) Remove material:

drill, grind, mill, ream

2) Add material:

flame, insulate, remetal, retread

3）Attach material:

bond, braze, crimp, rivet, solder, weld

4）Change the mechanical strength, the structure, or physical properties of a material:

anneal, cure, freeze, heat-treat, magnetize, normalize

5）Change the surface finish of a material:

buff, burnish, passivate, plate, polish

6）Change the shape of a material:

cast, extrude, spin, stamp

Part V　Maintenance English Application Practice

Choose the best answer from the following choices.

(　) 1. The manufacturer may establish procedures to provide feedback (from operators and maintenance facilities who comply with a CAP) on unairworthy conditions (whether covered or not covered by the CAP) that were discovered when complying.

A. 制造商可制定程序，对查询 CAP 时发现的不适航的情况（无论是否包含在 CAP 中）提供建议（来自遵循 CAP 的营运人和维修厂）。

B. 制造商可制定程序，对遵循 CAP 时发现的不适航的情况（无论是否包含在 CAP 中）提供反馈（来自遵循 CAP 的营运人和维修厂）。

C. 制造商可制定程序，对补充 CAP 时发现的不适航的情况（无论是否包含在 CAP 中）提供方案（来自遵循 CAP 的营运人和维修厂）。

(　) 2. Get the approval of the manufacturer before you repair this unit.

A. 修理这个组件之前要得到制造商的许可。

B. 修理这个组件之前要得到航空公司的许可。

C. 修理这个组件之前要得到局方的许可。

(　) 3. Preparation of technical data for certain chapters in the maintenance manual require joint contribution and close coordination between the manufacturer of the airframe and the manufacturers of various other components used in making the completed airplane.

A. 维修手册中某些章节的技术资料的编写，需要机身制造商和用于制造整架飞机的各种其他部件的制造商共同奉献和密切计划。

B. 维修手册中某些章节的技术资料的编写，需要机身制造商和整机的各种其他各制造商共同奉献和密切计划。

C. 维修手册中某些章节的技术资料的编写，需要机身制造商和用于制造整机的各种其他部件的制造商共同合作和密切协调。

(　) 4. Copies of the appropriate forms and a description of the entries to be completed are available to you from an ABC dealer or factory customer service representative.

A. 可以从 ABC 经销商或工厂客户服务代表处获得相应表格的副本和需要填写条目

的说明。

B．ABC 经销商或工厂客户服务代表处需获取相应表格的副本和需要填写条目的说明。

C．ABC 经销商或工厂客户服务代表处必须提供相应表格的副本和需要填写条目的说明。

(　) 5．The manufacturer shall include，wherever applicable，the office department to contact to obtain special repair or overhaul information not covered in the maintenance manual.

A．在适用的情况下，制造商应包含相关办公室或部门，以便获取维护手册中未涵盖的特殊修理或维修信息。

B．在适用的情况下，制造商应该联系相关办公室或部门，以便获取维护手册中未涵盖的特殊修理或翻修信息。

C．在适用的情况下，制造商应该有与之联系的办公室或部门，以便获取维护手册中未涵盖的特殊修理或大修信息。

(　) 6．For the CAP to be successful on a continuing basis, it is essential that a free flow of information exists between the operator，FAA and ABC Aircraft Company.

A．为了使 CAP 持续取得成功，运营商、FAA 和 ABC 飞机公司之间信息自由流通至关重要。

B．为了在 CAP 基础上持续取得成功，运营商、FAA 和 ABC 飞机公司之间信息自由流通至关重要。

C．为了使 CAP 取得成功的基础，运营商、FAA 和 ABC 飞机公司之间的信息自由流通是至关重要的。

(　) 7．To calibrate the test set，refer to the manufacturer’s instructions.

A．要校正测试设置，参考制造商说明。

B．要修改测试结果，需要征得制造商许可。

C．要调整测试结果，参考制造商说明。

(　) 8．制造商在本节中应提供一份设备清单，作为单独的文件附于本手册，或单独提供。

A．In this section which is a separate attachment of the handbook or as a separate document，the manufacturer shall provide an equipment list.

B．The manufacturer shall provide an equipment list in this section，in a separate document appended to the handbook，or as a separate document.

C．Being a separate appendix of the handbook or as a separate document，the manufacturer shall provide a list of equipment of this section.

(　) 9．ASD、简化技术英语维护集团（STEMG）以及与 STE 生产相关的任何组织或公司均无意或暗示对提供培训或生产支持性软件产品的组织、公司和个人提供任何担保或认可。

A．Neither ASD，the Simplified Technical English Maintenance Group（STEMG），nor any organization or company associated with the production of STE intend or imply any warranty or endorsement to these organizations, companies, and individuals that give training or produce supporting software products.

B. ASD, the Simplified Technical English Maintenance Group (STEMG), and any organization or company associated with the production of STE, which intend or imply any warranty or endorsement to these organizations, companies, and individuals will not give training or produce supporting software products.

C. Neither ASD, or any organization or company associated with the production of STE intend or, imply any warranty or endorsement to these organizations, companies, and individuals that give training or produce supporting software products.

(　) 10. 无论使用哪种方法，任何符合 GAMA 格式的出版物都将使用相同的基本编号系统。

A. No matter which method is adopted, the form of any publication shall comply with GAMA and use the identical basic numbering system.

B. Regardless of method used, any publication conforming to the GAMA format will use the same basic numbering system.

C. In spite of method adopted, any publication will adopt the same basic numbering system complying with the GAMA format.

(　) 11. 制造商提供用于缩微胶卷的材料，包括修订版，应仅在单面以适合拍摄的形式提供。

A. Material supported by the manufacturer for microfilming, including removals, shall be given on one edge only in a form suitable for photographing.

B. Material required the manufacturer for microfilming, including adjustments, shall be provided on one edge only in the format of suitable for photographing.

C. Material supplied by the manufacturer for microfilming, including revisions, shall be provided on one side only in a form suitable for photographing.

(　) 12. The listing of CAP Inspections contains all continuing airworthiness program inspections in the CAP, in numerical order by CAP inspection number, with full title, date, and effectivity and inspection compliance.

A. In the listing of CAP Inspections, apart from full title, date, effectivity and inspection compliance, all continuing airworthiness program Inspections in the CAP are excluded in numerical order on the basis of CAP inspection number.

B. Along with full title, date, effectivity and inspection compliance, all continuing airworthiness program inspections in the CAP are excluded in numerical order in the light of CAP inspection number, in the Listing of CAP Inspections.

C. All continuing airworthiness program inspections in the CAP are covered, in numerical order, by CAP inspection number, in the listing of CAP inspections along with full title, date, and effectivity and inspection compliance.

(　) 13. This specification was developed by the general aviation manufacturers association as guidance for airplane, engine and component manufacturers in preparing continuing airworthiness program inspection documents (which may also be called Continuing Airworthiness Programs or CAPs). CAPs may be prepared for all types of general aviation airplanes and components certificated under the applicable federal aviation regulations.

A. When the airplane, engine and component manufacturers prepare continuing airworthiness program inspection documents (which may also be called Continuing Airworthiness Programs or CAPs), they can consult this specification as a guidance.

B. General aviation manufacturers association built up this specification to guide the airplane engine and component manufacturers to prepare continuing airworthiness program inspection documents (which may also be called Continuing Airworthiness Programs or CAPs). All types of general aviation airplanes and components certified under the applicable federal aviation regulations may be utilized to prepare CAPs.

C. General aviation manufacturers association built up this specification to guide the airplane engine and component manufacturers to prepare continuing airworthiness program inspection documents (which may also be called Continuing Airworthiness Programs or CAPs). For all type of general aviation airplanes and components certified under the applicable federal aviation regulations, CAPs may be prepared.

(　) 14. It is expected that the manufacturers of major components will provide maintenance, troubleshooting, inspection and repair data in the format of this specification.

A. The manufacturers of major components must provide maintenance, troubleshooting, inspection and repair data in the same layout of this specification.

B. The manufacturers of major components is looking forward to gain maintenance, troubleshooting, inspection and repair data with the same form of this specification.

C. Manufacturers of major components should provide maintenance, troubleshooting, inspection and repair data with the same format of this specification.

(　) 15. In some cases, the specification is used to prepare for continuing airworthiness program for acceptance by another airworthiness authority other than FAA.

A. Sometimes FAA and other airworthiness authority may accept the continuing airworthiness program that uses the specification.

B. Sometimes we use the specification to prepare a continuing airworthiness program that will be accepted by non-FAA airworthiness authority.

C. Sometimes during preparing a continuing airworthiness program, using the specification may be accepted by FAA and other airworthiness authority.

Part VI Safety and Regulations

Craftsman Spirit

Craftsman Spirit

The Chinese Nation has always had a fine tradition of diligence, courage and unremitting self-improvement, and the concept of hard work, honest work and creative work.

In September 2021, the CPC Central Committee approved the first batch of great spirits sorted out by the Publicity Department of the CPC Central Committee to be included in The Spirit of the Chinese Nation, including the spirit of labor.What is the spirit of labor in the new era?

Labor spirit is the attitude, concept and spirit that every worker holds in the working process to create a better life.A great era needs great spirit, and great spirit comes from great people. With the economic and social development, the way of labor is changing, and the new era also endows the spirit of labor with new connotation.On 24 November 2020, General Secretary Xi Jinping pointed out at the National Conference on Model Workers and Advanced Workers, in the long-term practice, we have cultivated the spirit of labor, which is about working with pride, enthusiasm, diligence, and honesty.

General Secretary Xi Jinping's series of important speeches on labor and labor spirit provide an important basis for us to correctly understand labor spirit, pointing out the direction for upholding and carrying forward the work spirit in the new era. " Working with pride, enthusiasm " means that we should love and devote ourselves to our work, growing up on the basis of our posts, reflecting our value, showing our style and feeling happy in our work. We should be proud of hard work, and be shamed in indolence.

" Love what you do " .Xue Ying is an ordinary worker in AVIC Xi'an Aircraft Industry Group Co. LTD. for 29 years. She and her colleagues worked hard every day and delivered more than 7,000 high-quality vertical fins over the years.It is the spirit of labor in thousands of Xue Ying, their willing to learn, their willing to do and their willing to study, has pushed Chinese manufacturing to a new height.

" Working with diligence, and honesty" , means it is necessary to work hard, no matter how the times develop, diligence and honesty are always the nature of labor.Whether it is front-line medical workers or community workers, it is the hard work and dedication of workers from all walks of life that has brought together a powerful force to defeat the epidemic.

As General Secretary Xi Jinping has pointed out, " Socialism has been achieved through hard work, and the new era has been achieved through hard work." Tens of millions of workers are working hard in different jobs, and their diligence and hard work have instilled a working spirit.

At present, it is urgent to speed up the building of a large army of knowledge-based, skilled and innovative workers. Labor spirit cultivation is the inevitable requirement of training and bringing up the new talent of the times. The new era needs high-quality workers who love work, work hard and are good at work.

Have you got it? Labor has created the Chinese nation, created a glorious history of the Chinese nation, and will certainly create a bright future of the Chinese nation. We should learn and carry forward the spirit of labor.

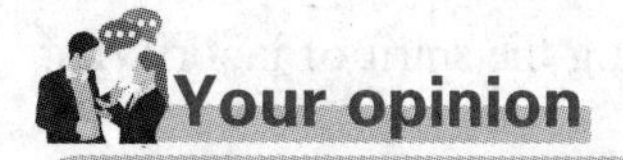

Your opinion

The labor spirit is a long tradition of Chinese nation, what is the connotation of labor spirit in the new era?

Keys for Reference

译文

Module

03

Category of Aircraft Maintenance Manuals

Project 1 Category of Aircraft Maintenance Manuals

Objectives

To know the category of aircraft maintenance manual.

To understand the typical contents of aircraft maintenance manual.

To master the usage of holding tools.

To know the latest manual for aircraft maintenance personnel.

Part Ⅰ Aviation Situational Communication

1. Read aloud the topic-related sentences for aviation communication and try to match them with their Chinese meaning.

Common expressions in emergency communications.

_______ (1) We cannot extend the left landing gear. Request foaming the runway.

_______ (2) We made the go-around due to GPS error.

_______ (3) We have a problem with the aircraft battery. Request priority landing.

_______ (4) Our operations manual forbids this procedure.

________ （5）We have lost electrical power to the cabin air compressor.

________ （6）The air conditioning system has malfunctioned.

________ （7）We have a problem with fuel temperature.

________ （8）One of our generators has failed.

Listen

A．我们飞机电瓶有问题。申请优先落地。

B．我们不能放出左起落架，申请在跑道上喷洒泡沫。

C．空调系统已故障。

D．我们的燃油温度有问题。

E．我们一部发电机故障了。

F．我们的运行手册禁止该程序。

G．我们复飞是因为 GPS 误差。

H．我们的客舱空气压缩机失去电源。

2. Listen to some sentences for aviation communication carefully and try to fill in the blanks.

（1）Our pressurization system has ________.

（2）Our pitch ________ has malfunctioned. We cannot move the elevators.

（3）We made the ________ because our landing gear would not extent normally.

（4）We have a problem with fuel temperature. Request holding position on the ________ .

（5）One of our generators has ________ . At present moment，we are still able to continue to our destination.

Listen

Part II Aviation Reading

2.1 Aviation Reading

Pre-reading Questions

1. According to the article, how many categories can aircraft maintenance manuals be divided into? What are they?

____

2. In order to reduce the confusion of maintenance manuals, which organization has established uniform standards for maintenance manuals?

____

3. How often should A320 AMM be revised regularly according to ATA 2200 specification?

____

Category of Aircraft Maintenance Manuals

Maintenance manual system refers to a set of manuals related to the maintenance/repair of the aircraft and accessories. The manuals provide working procedures, standards, instructions and other relevant information during maintenance to ensure that the aircraft is continuously airworthy through qualified maintenance.

Because of different manufacturers and aircraft's models, maintenance manuals are quite different. In an effort to reduce airline maintenance confusion, the Air Transport Association of America (ATA) has developed uniform standards for maintenance manuals for coordination among all aircraft manufacturers. The manual is written and published in accordance with ATA 2200 technical specification.

There are numerous manuals on aircraft repair and maintenance. Different documents provide different assistance for repair or maintenance activities. Manuals are usually divided into the following categories.

(1) Customization manuals: The information contained in the manual is only applicable to a specific customer or airline, usually on the CD or content with the customer code, and the validity is expressed by FSN (Fleet Serial Number). The customization manual is dedicated to a particular customer's fleet. Common customizations manuals include:

FCOM (Flight Crew Operations Manual) provides the flight crew with information on operating limits, operating procedures, aircraft performance, and aircraft systems required to operate an aircraft safely and efficiently.

MMEL (Master Minimum Equipment List) is a list of minimum equipment prepared by an aircraft operator for its operation, taking into account the configuration, operating procedures and conditions of the aircraft, which is a customized document, and the minimum equipment list (MEL) of different companies cannot be used universally.

TSM (Trouble Shooting Manual) provides maintenance personnel with recommended fault isolation procedures for troubleshooting.

IPC (Illustrated Parts Catalog) includes the part numbers of all parts on the aircraft, which is used by route maintenance personnel to identify, confirm, replace, and locate standard parts on the aircraft, and is used by the aviation materials department to order, store and issue materials.

ASM (Aircraft Schematics Manual) shows the working principle, system configuration, function, circuit operation and logical relationship of electronic/electrical systems in the form of system block diagram, system schematic diagram and system simplified schematic diagram. It helps maintenance personnel understand the system and troubleshoot faults.

AWM (Aircraft Wiring Manual) is used to understand, operate, troubleshoot and maintain electronic and electrical systems installed on aircraft.

(2) Aircraft manuals: The information contained in the manual is applicable to one or more aircraft of all airlines, usually on CD or on the content of the aircraft, the validity of the MSN (Manufacturing Serial Number). The model manual is used for a certain model series, and the same model can be used by all customers. Common model manuals include:

SRM (Structure Repair Manual) is formulated by aircraft manufacturers according to the type of aircraft and airworthiness requirements. It is the main basis for determining the maintenance and repair plan of aircraft structure.

NTM (Non-destructive Testing Manual) is nondestructive testing procedures developed by airframe, engine and component manufacturers.

ARM (Aircraft Recovery Manual) is an important reference document for airlines and aviation authorities to make material and technical preparations for damaged aircraft and implement rescue of damaged aircraft.

(3) General manuals: The manual information is applicable to all aircraft types of all airlines. The common manual is used for all aircraft types, such as airbus ESPM/PMS can be used for all aircraft of airbus. Common general manuals include:

ESPM (Electrical Standard Practices Manual) provides a quick and easy way to find detailed descriptions, construction procedures, construction tools, and other information about aircraft electronic and electrical connection components (contacts - pins, jacks, terminals, splices etc.).

CML (Consumable Materials Lists) is a consumable material list. It is mainly used to find some consumable parts, such as sealant, curing glue, structural paint, grease, cleaning agent etc.

PMS (Process Material Specification) provides information about manufacturing process and materials of aircraft construction, which is used in connection with the design drawings.

ATA 2200 technical specification has clear requirements for the revision of the manual, and the commonly used manual should be revised regularly. Manuals for aircraft manufacturers and engine manufacturers are usually revised periodically, such as A320 AMM/IPC/TSM/AWM/SRM every 3 months.

Words and Expressions

category [ˈkætəgərɪ] *n.* 种类，类别

accessory [əkˈsesərɪ] *n.* 配件

customization [ˈkʌstəmaɪzeɪʃn] *n.* 用户化；定制化

validity [vəˈlɪdəti] *n.* 合法性；有效性

revision [rɪˈvɪʒən] *n.* 修改；校订版

FCOM (Flight Crew Operations Manual) 飞行组使用手册

MMEL (Master Minimum Equipment List) 主最低设备清单

AMM (Aircraft Maintenance Manual) 飞机维修手册

TSM (Trouble Shooting Manual) 故障分析手册

IPC (Illustrated Parts Catalog) 零部件图解目录

ASM (Aircraft Schematics Manual) 飞机原理线路图手册

AWM (Aircraft Wiring Manual) 飞机线路图手册

SRM (Structure Repair Manual) 结构修理手册

NTM (Non-destructive Testing Manual) 无损探伤手册

ARM (Aircraft Recovery Manual) 飞机恢复手册

ESPM (Electrical Standard Practices Manual) 电气标准施工手册

SM (Standard Manual) 标准手册

CML (Consumable Materials Lists) 消耗性材料清单

PMS (Process Material Specification) 过程材料规范

Fill in the blanks with the proper words given below, changing the form if necessary.

IPC	customization	TSM	SRM	AWM

1. The _______ has now become a rule.

2. The _________ manual is an important reference for troubleshooting.

3. _________ provide the identification and typical repair information of vulnerable structural parts within the allowable damage range.

4. _________ provides detailed circuit connections.

5. ___________ provides an illustrated list of route replacement parts（LRU）and detailed part number information.

2.2 Manual Reading

There are some tasks for you to fulfill. You should read the Manual Reading materials carefully and then finish the tasks.

Maintenance manuals provided by aircraft and engine manufacturers that include instructions for maintaining all systems and functional components installed in aircraft.

The content of the Aircraft Maintenance Manual AMM is used to meet the field personnel to maintain installed on the aircraft component, system, structure of data, rather than overhaul and maintenance personnel use the data of parts.

Typical aircraft maintenance manual include:

(1) Description of each system;

(2) Lubrication instructions, refueling times, lubricating grease and oil used in different systems;

(3) Pressure and electrical load in different systems;

(4) Tolerances for the normal operation of the aircraft, and necessary adjustments;

(5) Methods of leveling, jacking and towing aircraft;

(6) Balancing control surface method;

(7) Inspection intervals and scope required by the aircraft in normal operation;

(8) Simple structure inspection and maintenance methods of aircraft;

(9) General visual and perforation inspection techniques;

(10) Special work orders allowed for various outfields.

Words for Reference

lubrication	[ˌluːbrɪ'keɪʃn]	*n.* 润滑
tolerance	['tɒlərəns]	*n.* 容许偏差
leveling	['levəlɪŋ]	*n.* 水准测量；校平
jacking	['dʒækɪŋ]	*n.* 顶托
towing	['təʊɪŋ]	*n.* 牵引支架
interval	['ɪntəvl]	*n.* 间隔
borehole	['bɔːˌhəʊl]	*n.* 钻孔
perforation	[pɜːfə'reɪʃn]	*n.* 穿孔
outfield	['aʊtfiːld]	*n.* 外场

Reading Tasks

1. Judge the following statements according to the contents of the AMM.

(1) Each aircraft and engine manufacturer provides maintenance manuals. (　)

(2) Maintenance manuals contain instructions for maintaining all systems and functional components installed in aircraft. (　)

(3) Aircraft maintenance manuals are not only suitable for field maintenance personnel, but also for component maintenance personnel. (　)

2. Fill in the blanks according to the above reading materials.

Typical aircraft maintenance manual include: ________ of each system, ________, refueling times, lubricating grease and oil used in different systems: ________in different systems, tolerances for the normal operation of the aircraft, and necessary adjustments; methods of leveling, jacking and towing aircraft; balancing control surface method ect.

Translate the following sentences into Chinese.

1. Lubrication instructions, refueling times, lubricating grease and oil used in different systems.

____

2. Pressure and electrical load in different systems.

____

3. Tolerances for the normal operation of the aircraft, and necessary adjustments.

____

4. Methods of leveling, jacking and towing aircraft.

____

5. Balancing control surface method.

____

Translate the following abbreviations into corresponding Chinese.

1. FCOM ____________
2. TSM ____________
3. IPC____________
4. SRM____________
5. NTM____________

Part III Tools and Equipments

Holding Tools

1. Match the words or phrases with the translations.

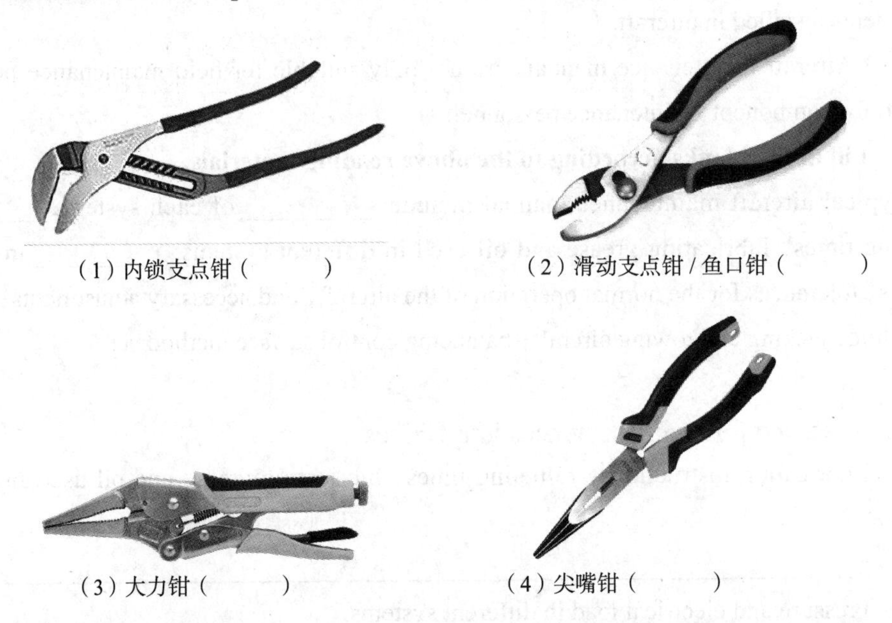

(1) 内锁支点钳（　　）

(2) 滑动支点钳/鱼口钳（　　）

(3) 大力钳（　　）

(4) 尖嘴钳（　　）

A. Slip joint plier	B. Needle nose plier
C. Interlocking joint plier	D. Locking plier

2. Fill in the words or phrases in the blanks.

Slip joint plier	Needle nose plier
Interlocking joint plier	Locking plier

______

(1) It is also called water pipe plier or eagle nose plier. The plier mouth can slide parallel along the chute, and the clamping range can be adjusted. The clamping range is large, and the clamping force is large, which can be used to clamp sealing nuts, pipe connections and special shaped parts.

______

(2) It is a special set of high leverage plier, which is adjustable and may be locked into place by a specially designed toggle in the handles. There are three common jaws of the pliers: pointed, round and flat, and they are often used to remove the deformed screws and the bolts with broken heads higher than the workpiece.

(3) It is also called fish plier. The length of the pliers is 150–200 mm in different sizes, and there is a double hole slot in its hinge point. By sliding the position of the fulcrum in the double hole, the clamping range can be changed, and the fuse can be cut by its root. Do not clamp fasteners during maintenance, which is easy to cause permanent damage to the shape of aluminum and copper nuts.

(4) They are used to hold small objects. There are two kinds of mouth shape, straight and oblique. They can be used for narrow space operation, and for fuse construction, etc.

Part IV Simplified Technical English

Noun Clusters

名词词组

★主要规则：使用名词词组时，不要超过三个单词。

技术文本通常包含单词词组，这些单词词组通常由名词、形容词组成，在句子中做主语或宾语。名词词组中，单词越多越不易于理解，并产生歧义，不超过三个单词会使理解更容易也更准确。

Please judge which of the following sentences is allowed by STE.

(1) Remove the bolts that attach the transmission housing to the engine.

Remove the engine transmission housing attachment bolts.

(2) Runway light connection resistance calibration.

Calibration of the resistance of the runway light connection.

Keys:

Non-STE: Remove the engine transmission housing attachment bolts.

STE: Remove the bolts that attach the transmission housing to the engine.

Non-STE: Runway light connection resistance calibration.

STE: Calibration of the resistance of the runway light connection.

Part V Maintenance English Application Practice

Choose the best answer from the following choices.

( ) 1. Definitions should be worded as simply as possible and must conform to the use of the defined terms in the handbook.

A. 定义的措辞应尽可能简单，并且必须与手册中定义的术语相符。

B．定义的措辞必须与手册中使用的定义术语一致，还应以简单的方式来表达。

C．定义的措辞应简单并符合手册中的术语。

（　）2．Except for fold out pages and start of chapters，all pages shall be printed on both sides.

A．包括折页和章节开头页，所有页面均应单面打印。

B．包括折页和章节开头，所有页面均应单面打印。

C．除了折页和章节开头页，所有页面均应双面打印。

（　）3．However，the title and applicable airplane or airplane series designation will be prominently shown on the cover or spine of the handbook.

A．但是，标题和适用的飞机或飞机系列名称要在手册的封面或书背上加粗显示。

B．但是，标题和适用的飞机或飞机系列名称要在手册的封面或书背上突出显示。

C．但是，标题和适用的飞机或飞机系列名称要在手册的封面或书脊上突出显示。

（　）4．The introduction section of the maintenance manual should contain either a list of vendor publications or refer to the location of such lists whether they are included in system descriptions or provided in separate documents.

A．维护手册的简介部分应包含供应商出版物的清单，或提及此类清单的位置，无论这些清单是包含在系统说明中还是在单独的文档中提供的。

B．维护手册的简介部分应包含供应商出版物的清单，并提及此类清单的位置，无论这些清单是包含在系统说明中还是在单独的文档中提供的。

C．维护手册的简介部分既不包含供应商出版物的清单，也未提及此类清单的位置，无论这些清单是包含在系统说明中还是在单独的文档中提供的。

（　）5．累积修订（包括备用页）占比不到手册总页数的60%时，无须重新编写手册。

A．When total revisions including left pages，above 60% of the total pages in the manual，a reassure of the manual should not be prepared.

B．When accumulated revisions including back-up pages，less than 60% of the total pages in the manual，a reissue of the manual should not be prepared.

C．When accumulated revisions including left pages，less than 60% of the total pages in the manual，a revise of the manual should not be prepared.

（　）6．本手册包括联邦航空法规要求向飞行员提供的材料和制造商提供的附加信息，并构成联邦航空管理局批准的飞机飞行手册。

A．This handbook includes the documents required to be finished to the pilot by the federal aviation regulations and additional information approved by the operator and contains the FAA approved airplane flight manual.

B．This handbook includes the material required to be furnished to the pilot by the federal aviation regulations and additional information provided by the manufacturer and constitutes the FAA approved airplane flight manual.

C．This handbook includes requirements of federal aviation regulations，material provided to pilot and accurate information provided by the manufactures，and establishes the FAA approved airplane flight manual.

(　) 7. 本规范编写时遵循的构成规则与 FAA 编写规则时用到的大体相同。

A. When making this specification, the rules are generally refer to the rules that used by FAA in making its rules.

B. When making this specification, we use the exact identical construction rules with FAA in making its specifications.

C. The rules of construction followed in the preparation of this specification are the same as generally used by the FAA in the preparation of its rules.

(　) 8. 本手册每一页不应使用实际的发行日期，而应使用“初始版本”一词。

A. Instead of using the actual date of issue on each page of an original issue of a handbook, the words “Original issue” may be used.

B. The actual release date should not be used on each page of this handbook. The term “Ahead release” should instead be used.

C. In place of using the actual date of release on each page of an original issue of a handbook, the letters “Original issue” have to be used.

(　) 9. 定义手册中使用的任何特殊术语，重点是那些可能被滥用或误解的术语。

A. Define any special grammars used in the handbook with important on those which could be missioned or misunderstood.

B. Define any special terminologies used in the handbook with emphasis on those which could be misused or misunderstood.

C. Identify any special terms used in the handbook with the point on those that could be missed or misunderstood.

(　) 10. The sequence of topics in the handbook is intended to increase in-flight usefulness.

A. The order of topics in the handbook is intended to increase in-flight usefulness.

B. The sequence of topics in the handbook is intended to increase in-flight useless.

C. The sequence of topics in the handbook is pretended to increase in-flight usefulness.

Part VI Safety and Regulations

In 2021, the International Civil Aviation Organization (ICAO) released the Competency-based Training and Assessment Manual for Aircraft Maintenance Personnel (Doc 10098) (2021 first edition), which is the first manual specifically prepared for aircraft maintenance personnel training. The manual consists of 6 chapters, which details the specific practices of competency assessment and training for aircraft maintenance personnel, mainly including the design and implementation guidelines of competency assessment and training for aircraft maintenance personnel, competency assessment and training for aviation maintenance personnel refresher training, license and privileges, etc. Guidance is provided to approved training institutions, approved maintenance institutions and national civil aviation authorities on how to determine the competency of aircraft maintenance personnel.

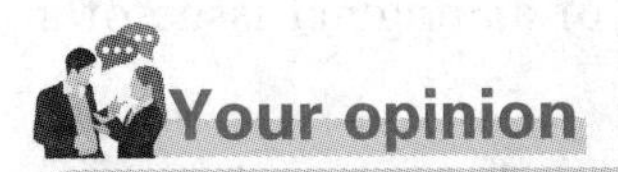

Your opinion

What chapters are included in the Aircraft Maintenance Personnel Assessment Manual (Doc 10098)(2021 1st Edition) issued by ICAO? And how to become a qualified maintenance personnel?

Project 2 Aircraft Maintenance Manual

Objectives

To know the Aircraft Maintenance Manual (AMM).

To understand the repair of flexible hose.

To master the usage of holding tools.

To perceive the civil aviation spirits "three reverences".

Part I Aviation Situational Communication

1. Read aloud the topic-related sentences for aviation communication and try to match them with their Chinese meaning.

Listen

_______ (1) The travel ranges of the outflow valves are clear.

_______ (2) Movement of the outflow valve can cause damage to equipment.

_______ (3) Parts shall be cleaned in a ventilated room.

_______ (4) Cleaning steps shall in no case be performed in the vicinity of any open flame.

_______ (5) Parts shall be protected by means of protective caps.

A．清洁程序在任何情况下都不能在明火旁进行。
B．外流阀的运动区域无障碍物。
C．部件应该用防护堵盖进行防护。
D．部件应在通风的房间内进行清洁。
E．外流阀的移动可能导致设备损坏。

2. Listen to some short dialogues between A and B. After the dialogues there are some questions. Please choose the most appropriate answer from the choices.

（1）Situation one

1．What's the meaning of "clear"?（　　）

A．允许　　C．无障碍的

B．清除　　D．没有

2．What's meaning of "ambient"?（　　）

A．琥珀色的　　C．安培

B．外界的，环境的　　D．伏特

3．What's meaning of "pressurization"?（　　）

A．增压　　C．减压

B．负压　　D．正压

（2）Situation two

1．Where shall parts be cleaned?（　　）

A．in a spare room.　　C．in a sealed room.

B．in a vacuum room.　　D．in a duly ventilated room.

2．What's meaning of "in the vicinity of"?（　　）

A．与……相关　　C．关于

B．在……附近　　D．涉及

Listen

Part II　Aviation Reading

2.1　Aviation Reading

1．What is AMM?

____

2．How many parts does AMM contain? And what are they?

____

3. What is ASN?

Introduction of Aircraft Maintenance Manual

Aircraft Maintenance Manual (AMM) is the Publication, which has been prepared by the Boeing Commercial Airplane Group in accordance with Air Transport Association of America Specification No. 100 (ATA-100). This is the specification for manufacturer's technical data.

ATA Numbering System

Aircraft Maintenance Manual (AMM) divides into two parts: the System Description Section (SDS) and the Practices and Procedures (PP). The SDS is the publication of part one of AMM, which contains the functions, operation, configuration and control of the airplane system and subsystems. It explains how the systems and components operate, and illustrates how systems are constructed, and gives the descriptions of the relationships between the interfaces and other components. The PP is the publication of part two of AMM, which describes the maintenance practices and procedures to do maintenance on the airplane.

In accordance with ATA-100, AMM provides procedures for both scheduled and un-scheduled maintenance. At the beginning of this manual, there is front matter for easy lookup. This material includes the title, effective aircraft, transmittal letter and revision lists, introduction, and service bulletin etc. What's more, the maintenance manual is divided into chapters and groups of chapters (Table 1). These represent a functional break-down of the airplane and its system.

Table 1 Groups of Chapter

Numbers of Chapters	Names of groups
Chapter 1-19	Aircraft General
Chapter 20-49	Airframe Systems
Chapter 50-59	Structures
Chapter 60-89	Power plant

According to ATA numbering system, there is athree elements number XX-YY-ZZ in each page of AMM for identifying a single functional item within the wider scope of each chapter, which is known as the assigned subject number (ASN). The three elements of the indicator each contain two digits (Fig. 3-1). The first element XXmeans the chapter or system number allocated by ATA specification No. 100. The second element YY refers to a section or sub-system of this chapter. The first digit of the second element is allocated by ATA specification No.100, and the second digit is designated by manufacturers, which is prone to be different from each other. The third element ZZ indicates a subject or a unit, which is level breakdown allocated by manufacturers. For examples 27 indicates flight control system, 31 means elevator & tap control section or sub-system; 11 refers a subject or unit number, that is elevator installation/removal/inspection/check operation.

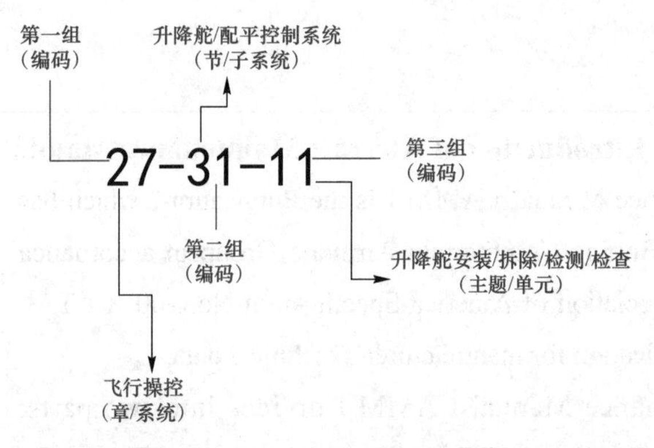

Fig. 3-1 Chapter numbering

Besides the ASN, there is another number in the lower right corner, which is page block. PB is short for Page Block. In order to facilitate reference and management, the specific topics in the sections of the maintenance manual are indicated in the form of page blocks. The page block covers all the operation steps in aircraft maintenance or repair work. For examples: The information of description and operation of system, components is found PB 001; any standard maintenance practices are in PB 201; any task related to service, e.g. replenishment or lubricating is presented in PB 301 etc.

Words and Expressions

AMM (Aircraft Maintenance Manual) 飞机维修手册
ATA-100 (Air Transport Association of America Specification No. 100) 美国航空运输协会规范第 100 号
SDS (Systems Description Section) 系统说明部分
PP (Practices and Procedures) 实践与程序
ASN (Assigned Subject Number) 指定功能号
front matter 前言
scheduled maintenance 定检
transmittal letter 手册发送说明
service bulletin 服务通告
page block 页面块
configuration [kənˌfɪgə'reɪʃn] *n.* 布局，构造；配置
effective [ɪ'fektɪv] *adj.* 有效的
breakdown ['breɪkdaʊn] *n.* 分类，细目列表

Fill in the blanks with the proper words given below, changing the form if necessary.

AMM	page block	front matter	configuration	scheduled maintenance

1. The ________ covers all the operation steps in aircraft maintenance or repair work.

2. The ________ of AMM includes the title, effective aircraft, transmittal letter and revision lists, introduction, and service bulletin etc.

3. The aircraft had undergone its last ________ on November 16, 2022.

4. This ________ does not deliver a system without frictions.

5. ________ is the specification for manufacturer's technical data, which has been prepared by the Boeing Commercial Airplane Group.

2.2 Manual Reading

There are some tasks for you to fulfill. You should read the Manual Reading materials carefully and finish the tasks.

AMM manual is the most important manual in the maintenance manual system. Pay attention to the validity of the content in the process of using the manual, read the "Warning", "Caution" and "Notes" information of the task before work, and complete the maintenance/repair work in strict accordance with the task.

A WARNING gives a condition or tells personnel what part of an operation or maintenance procedure, which if not obeyed, can cause injury or death.

A CAUTION gives a condition or tells personnel what part of an operation or maintenance procedure, which if not obeyed, can cause damage to the equipment.

A NOTE gives data, not commands. The note helps personnel when they do the relate instruction.

Warnings and cautions go before the applicable paragraph or step. Notes follow the applicable paragraph or step.

Flexible Hose−Repairs

General

A. This procedure contains one task. The task gives instructions for repair of hydraulic, pneumatic, water, and other tube assemblies. The repair is replacement of unserviceable rigid tubes with flexible hoses as a temporary repair.

B. If there are other instruction on other procedures or manuals for a specified system, obey those instructions.

CAUTION: PUT CAPS ON HOSES AND FITTINGS WHEN THEY ARE NOT CONNECTED. IF YOU DO NOT USE CAPS, CONTAMINATION OF THE HOSES, AND DAMAGE TO THE SYSTEM COMPONENTS AND LEAKAGE OF HYDRAULIC FLUID COULD OCCUR.YOU MUST REMOVE ALL HYDRAULIC FLUID THAT CAME OUT OF THE HOSE AND FELL ON THE AIRPLANE.HYDRAULIC FLUID CAN CAUSE CORROSION AND DAMAGE TO THE AIRPLANE.

C. Keep caps on hoses and connections to keep moisture or other contamination out of the system until the hose is connected again.

Selected from AMM

Words for Reference

hose	[həʊz]	*n.* 水管；橡皮软管
hydraulic	[haɪ'drɒlɪk]	*adj.* 液压的
pneumatic	[njʊ'mætɪk]	*adj.* 气动的
assembly	[ə'semblɪ]	*n.* 组装；装配；(待装配的) 成套部件
unserviceable	[ʌn'sɜːvɪsəbl]	*adj.* 不适用的
rigid	['rɪdʒɪd]	*adj.* 刚性的；坚硬的；不易弯曲的
cap	[kæp]	*n.* 盖；套；罩
contamination	[kənˌtæmɪ'neɪʃn]	*n.* 污染；污染物
leakage	['liːkɪdʒ]	*n.* 渗漏物；泄漏物
corrosion	[kə'rəʊʒn]	*n.* 腐蚀

Reading Tasks

1. Judge the following statements according to the contents of the AMM.

(1) You should read the " Warning", " Caution " and " Notes" information of the TASK after work. (   )

(2) The CAUTION helps personnel when they do the relate instruction. (   )

(3) Notes follow the applicable paragraph or step. (   )

2. Fill in the blanks according to the above reading materials.

(1) You should keep ______________ on hoses and fittings to keep contamination out of system before the hose is connected.

(2) If you do not use caps __________________, damage to the system components and __________________ could occur.

Translate the following sentences into Chinese.

1. Pay attention to the validity of the content in the process of using the manual.

____

2. Complete the maintenance/repair work in strict accordance with the task.

____

3. The task gives instruction for repair of hydraulic, pneumatic, water and other tube assemblies.

____

4. You must remove all hydraulic fluid that came out of the hose and fell on the airplane.

____

5. Put caps on hoses and fittings when they are not connected.

____

Translate the following abbreviations into corresponding Chinese.

1. AMM________________
2. ATA-100______________
3. SDS________________
4. PB_________________
5. ASN________________

Part III Tools and Equipments

Holding Tools

1. Match the words or phrases with the translations.

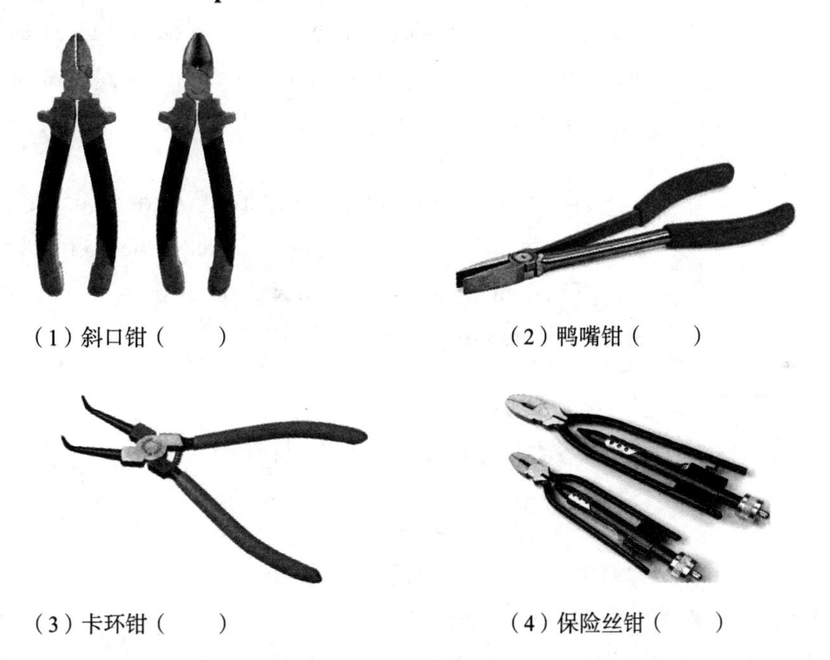

(1) 斜口钳（　　）
(2) 鸭嘴钳（　　）
(3) 卡环钳（　　）
(4) 保险丝钳（　　）

A. Convertible snap ring plier	B. Safety wire twister
C. Diagonal plier	D. Duckbill plier

2. Fill in the words or phrases in the blanks.

Diagonal plier	Duckbill plier
Convertible snap ring plier	Safety wire twister

(1) The head of plier has two convex tips, which can be divided into inner ring clamp and outer ring clamp. The inner clamp is used to shrink and deform the clamp ring to complete the disassembly and installation. External clamp is used to expand and deform the clamp ring to complete disassembly and installation.

(2) It is used for clamping, shearing and roration, which is a more efficient tool for fuse insurance of fasteners. However, due to the damage caused by the clamping of the jaws to the fuse clamping, it is easy to break, so it is prohibited to use its construction operation in some special areas, such as the elastic screw sleeve insurance of the steel cable.

(3) It is also known as wire pliers. The plier has a small angle to form a blade, which is used to cut metal wires, rivets, cotter pins, etc. It is often used in the construction of installation and removal fastener insurance.

(4) It has a flat mouth shaped like a duck beak. There are fine teeth in the mouth of the jaw to increase the friction against the grip. The longer handle provides good gripping capability and is used to tighten fuse knots.

Part IV Simplified Technical English

Verbs
动词

★规则：在程序性写作中只使用主动语态。

技术文本包括程序性写作（procedural writing）和描述性写作（descriptive writing）。在程序性写作和尽可能多的描述性写作中只使用主动语态。

Please change the following sentences from passive voice to active voice.

(1) The circuits are connected by a switching relay.

（当句中给出了施动者，通常是介词“by”的宾语时，将施动者放在句首用作主语。）

(2) Oil and grease are to be removed with a degreasing agent.

（在程序性写作中，动词改为祈使“命令”形式。）

(3) On the ground, the valve can be opened with the override handle.

（当句中没有给出施动者时，可以增加代词“you”或“we”作为主动语态的主语。）

Keys:

(1) A switching relay connects the circuits. (Active)

(2) Remove the oil and the grease with a degreasing agent. (Active)

(3) On the ground, you can open the valve with the override handle. (Active)

Part V Maintenance English Application Practice

Choose the best answer from the following choices.

( ) 1. Use of the word " CAUTION", with appropriate 10-point uppercase for the text, enclosed in a box, is recommended.

A. 建议使用"告诫"一词，并用适当的 10 号大写将文本括在一个框中。

B. 建议使用"注意"一词，并用适当的 10 号小写将文本括在一个框中。

C. 建议使用"警告"一词，并用适当的 10 号大写将文本括在一个框中。

( ) 2. Units of measurements shown in manual must be consistent.

A. 手册中所示的度量单位必须准确。

B. 手册中的度量单位必须一致。

C. 手册中的度量单位必须写明。

( ) 3. An appropriate warning or note shall be contained in each handbook to inform the operator that a current handbook is required to be in the airplane during flight and that it is the operator's responsibility to maintain the handbook in a current status.

A. 每本手册中均应包含适当的警告或注意事项，以告知飞行员在飞行过程中需要在飞机上携带现行有效的手册，并且营运人有责任将手册保持在现行有效状态。

B. 每本手册中均应包含适当的警告或笔记，以告知飞行员在飞行过程中需要在飞机上携带通用的手册，并且营运人有责任将手册保持在现行有效状态。

C. 每本手册中均应包含适当的警告或注意事项，以告知飞行员在飞行过程中需要在飞机上携带现行有效的手册，并且飞行员有责任将手册保持在现行有效状态。

( ) 4. As a general rule, a revision shall be prepared when a minor number of the total pages of the manual is affected.

A. 一般而言，当手册中小部分页数受影响时，受影响部分应准备修订。

B. 一般而言，当手册中有一小部分内容受到影响时，应准备修订。

C. 一般而言，当手册的总页数中有一小部分受到影响时，应准备修订。

( ) 5. Schematic diagrams may be used to indicate flow and to illustrate the operation of systems such as air control, electrical, fluid power, fuel and turbo systems.

A. 原理图可以用来表示流程，并说明系统的工作过程，如空气控制、电力、流体动力、燃油和压气机系统。

B. 原理图可以用来表示流程，说明系统的工作原理，如空气分配、电力、流体动力、燃油和涡轮系统。

C．原理图可以用来表示流程，并说明系统的操作，如空气控制、电力、流体动力、燃油和涡轮系统。

（ ）6．When an illustration is required to be reproduced horizontally on a page，the top of the illustration shall always be toward the left edge of the sheet.

A．当要求在页面上垂直复制插图时，插图的顶部应始终朝向纸张的左边缘。

B．当要求在页面上水平复制插图时，插图的顶部应始终朝向纸张的左边缘。

C．当要求在页面上垂直复制文本时，文本的顶部应始终朝向纸张的左边缘。

（ ）7．当原理图由于主题系统中的自动功能或相关控件而变得复杂时，这些特性应通过图中的解释性文本或随附文本或两者都有来指出。

A．Where schematic diagrams are complex by virtue of automatic features or articulated controls in the subject system，these characteristics should be pointed by means of explanatory text in the diagram or in the accompanying text or both.

B．Where schematic diagrams are complex by venture of automatic characters or articulated controls in the subject system，these characteristics should be pointed at by means of apparent text in the diagram，or in the accompanying text，or both.

C．Where schematic diagrams are complex by virtue of automatic features or interrelated controls in the subject system，these characteristics should be pointed out by means of explanatory text in the diagram or in the accompanying text，or both.

（ ）8．Warnings and cautions shall precede the text to which each applies，but notes may precede or follow applicable text depending on the material to be highlighted，warning，cautions and notes shall not contain procedural steps nor shall they be numbered.

A．Like warnings and cautions which must lines in the rear of the text，notes may precede or follow proper text up to the material to be emphasized. Warning，cautions and notes shall not provide procedural steps nor shall they be numbered.

B．Unlike warnings and cautions which must lines in the rear of the text，notes may precede or follow proper text up to the material to be emphasized. Warning cautions and notes shall not provide procedural steps nor shall they be numbered.

C．Proper warnings and cautions shall be provided ahead of the text while notes may precede or follow suitable text up to the material to be highlighted，warning，cautions and notes shall not show procedural steps nor shall they be numbered.

（ ）9．Each system/chapter/section will start with a new block of page numbers.

A．Each system/chapter/section will use a new block of page numbers as a beginning.

B．When you prepare each system/chapter/section you can continue to use the page numbers previously-used.

C．The beginning of each system/chapter/section shall list a new block of page numbers.

（ ）10．A revision means a modification of information in an existing manual.

A．A advisement of information in a ready-made manual can be taken as a revision.

B. A revision refers to an adjustment of information in a manual under publishing.

C. A modification of information in an existing manual can be called a revision.

( ) 11. Each chapter or section is to start with a general description of the systern and its operation and should be labeled description and operation.

A. Each chapter or section begins with either a general description of the system its operation or description and operation should be labeled.

B. Each chapter or section begins with neither a general description of the system nor its operation nor description and operation should be labeled.

C. Each chapter or section, which is to begin with both a general description of the system and its operation, should be named description and operation.

( ) 12. First second and third level heads and captions should be distinctive in size and/or style.

A. The first second level heads and captions should be exactly identical to the third level ones in size and/or style.

B. The first level heads and captions should maintain consistent with the second and third level ones.

C. The size and/or style of the first second and third level heads and captions should be different and easily noticed.

( ) 13. Arrows shall be used to indicate flow direction when needed to understand the schematic diagram.

A. When you need to understand a schematic diagram, use arrows to indicate flow orientation.

B. Arrows shall be used to signify flow direction when needed to realize the schematic form.

C. When need to comprehend the chart, you should use arrows point to the direction.

( ) 14. An appropriate warning or note shall be contained in each handbook to inform the operator that a current handbook is required to be in the airplane during flight and that it is the operator's responsibility to maintain the handbook in a current status.

A. An appropriate caution or note shall be contained in each handbook to inform the operator that a current handbook is required to be in the airplane during flight and that it is the operator's duty to keep the handbook in a current status.

B. A current handbook that an appropriate warning or note shall be obtained in each handbook to inform the operator is required to be in the airplane during flight and that it is the operator's responsibility to retain the handbook in a current conditions.

C. A suitable warning or note shall be involved in each handbook to inform the operator that a current handbook is required to be in the airplane during flight and that it is the operator's responsibility to maintain the handbook in a current status.

(   ) 15. Detailed information on more complex and specialized subjects, such as structural wiring diagrams or overhaul information, may either be included in the basic maintenance manual or made available as separate documents at the option of the manufacturer and to the extent the manufacturer determines appropriate.

A. Detailed information on more complex and specialized subjects such as structural wiring diagrams or overhaul information may be contained in both the basic maintenance manual and separate documents at the discretion of the manufacturer and to the degree the manufacturer determines appropriate.

B. The basic maintenance manual and separate documents may consist of such detailed information on more complex and special subjects as structural wiring diagrams or overhaul information at the discretion of the manufacturer and to the degree the manufacturer determines appropriate.

C. Such detailed information on more complex and specialized subjects as structural wiring diagrams or overhaul information may be contained in the basic maintenance manual or found in separate documents at the discretion of the manufacturer and to the degree the manufacturer determines appropriate.

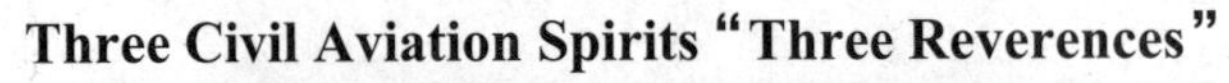

Part VI Safety and Regulations

Three Civil Aviation Spirits "Three Reverences"

General Secretary of the CPC Central Committee Xi Jinping put forward that as a civil aviation practitioner, we must deeply understand the true connotation of "Reverence for Life, Reverence for Regulations and Reverence for Duty".

Reverence for life reflects the value pursuit of the civil aviation industry. For aircraft maintenance work, to check and maintain every aircraft conscientiously and responsibly in daily work to ensure that the inspection and maintenance are in place. That is actually responsible for the life of every passenger, which is the best embodiment of reverence for life.

The ancients said "no rules, no square".Reverence for the rules reflects the operation of the civil aviation industry. In terms of daily work, it is required that every maintenance personnel must strictly abide by the rules and regulations.

Reverence for duty embodies the professional ethics of civil aviation personnel. At work, maintenance personnel should be strict with themselves and understand personal responsibility. No matter how boring it is, we should insist on it and take it seriously, because it is our job and our responsibility.

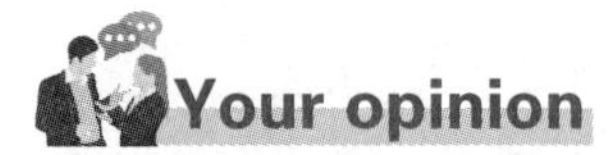

What are the "three reverences" of civil aviation spirit? How do we practice the spirit of "three reverences" in civil aviation?

Module

04 Typical Aircraft Structure

Project 1 Common Aircraft Structure

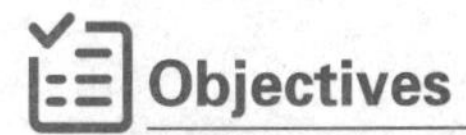

Objectives

To know the basic structure of an aircraft.

To understand the general description of the aircraft's structure and its fuselage.

To master the usage of files.

To perceive the "China Aviation Hero Crew".

Part I Aviation Situational Communication

1. Read aloud the topic-related sentences for aviation communication and try to match them with their Chinese meaning.

_______ (1) The wings are also used to store fuel and to install fuel system components.

_______ (2) The fuselage is an aircraft's main body section.
_______ (3) The vertical stabilizer is located at the top of the tail.
_______ (4) Power plant refers to the whole system that provides power for aircraft flying.
_______ (5) Landing gear is the main support part of an aircraft.

A. 动力装置是指为飞行器飞行提供动力的整个系统。
B. 机身是飞机的主体部分。
C. 机翼也用于储存燃料和安装燃料系统组件。
D. 垂直尾翼位于机尾顶部。
E. 起落架是飞机的主要支承部件。

Listen

2. **Listen to some aviation sentences carefully and try to fill in the blanks.**

(1) The fuselage is also used as the______ part of the whole body.
(2) The wings usually _______ of three main areas.
(3) The tail section of an _______ is empennage.
(4) Most _______ types of landing gear consists of three wheels.
(5) The _______ is the core component of the aircraft.

Listen

Part II Aviation Reading

2.1 Aviation Reading

Pre-reading Questions

1. What are the main structures of an aircraft?

____

2. How many main types of aircraft fuselage structure are there?

____

3. What is the rudder mainly used for?

____

The Basic Structure of an Aircraft

The Basic Structure of an Aircraft

Different types of aircraft can be designed for many different purposes, but for most of them, the main structures are almost the same, including fuselage, wing, empennage, landing gear and power plant (Fig. 4-1).

The fuselage is an aircraft's main body section, with tapered or rounded ends to make its shape aerodynamically smooth. Most fixed-wing aircraft have a single fuselage, often referred to as simply " the body (Fig. 4-2)" . The fuselage holds crew, passengers, cargo and equipment. There are two main types of aircraft fuselage structure: Monocoque structure and semi-monocoque structure. The monocoque structure uses a frame assembly and bulkheads to shape the fuselage, with the aircraft skin riveted directly to the frame. In the semi-monocoque structure, the skin is loaded by the longitudinal beam, the spacer and the stringers. This structure makes the fuselage more flexible and has a better strength and stiffness ratio. The fuselage consists of the forward nose, the forward body, the middle fuselage, the rear fuselage and the tail. The fuselage is also used as the central part of the whole body, which is assembled together with the wings, empennage, landing gear, power plant and so on to form the whole aircraft.

Fig. 4-1 Aircraft structure & components

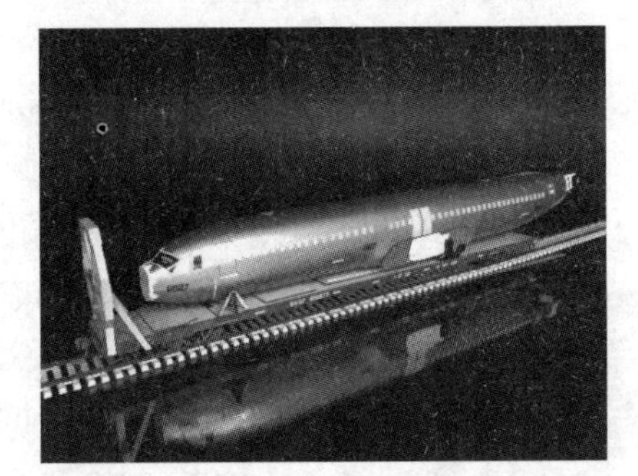

Fig. 4-2 Parts of an aircraft

Wings are an important part of an airplane (Fig. 4-3). They may be attached at the top, middle, or lower portion of the fuselage. Although they come in different location and shapes, their functions are the same: To generate lift and give the aircraft lateral stability and maneuverability. The wings can also be used to install landing gear, engines, flight controls and other systems. The wings are also used to store fuel and to install fuel system components. The wings usually consist of three main areas: the left wing, the center wing, and the right wing.

Fig. 4-3 Parts of an aircraft

The tail section of an airplane is empennage (Fig. 4-4). The empennage includes the entire tail group, consisting of fixed surfaces and movable surfaces. In general, the fixed surfaces are called the vertical stabilizer and the horizontal stabilizer. The vertical stabilizer is located at the top of the tail. The function of the horizontal stabilizer maintains longitudinal balance and provides pitch trim control. The movable surface includes the rudder, elevator, and one or more trim adjustment blades. The rudder is mainly used for yaw control.

Landing gear is the main support part of an aircraft when it is parked, taxiing, taking off or landing. Most common types of landing gear consists of three wheels—two main wheels and a third wheel positioned either at the front or rear of the airplane, usually at the front (Fig. 4–5). But the aircraft can also be fitted with pontoons for running on water, or sleds for skiing on snow.

Power plant refers to the whole system that provides power for aircraft flying, including the engine, propeller and other accessories. The most important part is the engine (Fig. 4–6). It is the core component of the aircraft. The structure of aircraft power plant is complex and self-contained and it is independent of the body of the aircraft.

Fig. 4–4 Parts of an aircraft

Fig. 4–5 Parts of an aircraft

Fig. 4–6 Parts of an aircraft

Words and Expressions

fuselage	[ˈfjuːzəlɑːʒ]	*n.* 机身（飞机）
empennage	[ˈempɪnɪdʒ]	*n.* 尾翼，尾部
landing gear		起落架；起落装置，着陆装置
power plant		发电厂；动力装置
aerodynamically	[ˌeərəʊdaɪˈnæmɪkli]	*adv.* 空气动力学地
monocoque	[ˈmɒnəˌkɒk]	*n.* 硬壳式构造
semi-monocoque	[ˈsemi ˈmɒnəˌkɒk]	*n.* 半硬壳式
rear	[rɪə(r)]	*n.* 后部，背部
lift	[lɪft]	*n.* （空气的）升力，提升力
stability	[stəˈbɪləti]	*n.* 稳定（性），稳固（性）；坚定，恒心
install	[ɪnˈstɔːl]	*v.* 安装，设置
engine	[ˈendʒɪn]	*n.* 发动机，引擎
vertical stabilizer		垂直安定面
horizontal stabilizer		水平安定面
pitch trim control		俯仰配平控制
rudder	[ˈrʌdə(r)]	*n.* 船舵；飞机方向舵
elevator	[ˈelɪveɪtə(r)]	*n.* （飞行器的）升降舵
yaw control		偏航控制
accessory	[əkˈsesəri]	*n.* 附件，配件

Fill in the blanks with the proper words given below, changing the form if necessary.

fuselage	empennage	landing gear	monocoque	accessory

1. On most passenger jets, the wings and ______ generate about 90% and 10% of the lift respectively.
2. Airplanes retract their ______ while in flight.
3. The ______ frame is made of hardwood and is exceptionally light weight.
4. Everything is suddenly now an ______ rather than an essential.
5. The moveable horizontal surface of the ______ is used for pitch trim.

2.2 Manual Reading

There are some tasks for you to fulfill. You should read the Manual Reading materials carefully and finish the tasks.

STRUCTURES—DESCRIPTION AND OPERATION

1. General

A. The structure of the airplane is designed to provide maximum strength with minimum weight. This object has been achieved by designing alternate load paths into the structure, so that a failure of one segment cannot endanger the airplane, and also by the use of appropriately selected materials. The materials most commonly used throughout the structure are aluminum, steel, and magnesium alloys. Of these, the most extensively used are certain aluminum alloys selected according to the particular type of load they are best suited to withstand.

B. Aluminum and fiberglass honeycomb core material is used extensively on secondary areas of structures and many of the flight surfaces.

C. Maintenance practices concerning blowout doors and panels are covered in applicable chapters on structures.

2. Fuselage

A. The fuselage is a semi-monocoque structure with the skin reinforced by circumferential frames and longitudinal stringers. It is composed of four sections: Body sections 41, 43, 46, and 48, of which the forward three together extend from body station 178 to body station (STA) 1016 and contain all the passenger, crew, and cargo accommodations. The fourth section of the fuselage is at the aft end and provides support for the empennage.

B. The entire shell of the fuselage between body stations 178 and 1016 is pressurized with the exception of the cavity enclosing the nose gear wheel well, and the large cutout which accommodates the center wing box and main landing gear (MLG) well. Structural continuity is provided across this latter area by a keel beam which passes beneath the center wing box and connects to a beam across the main landing gear wheel well. The whole pressurized portion of the fuselage is provided with a floor consisting of horizontal transverse beams attached to the fuselage frames and surmounted by longitudinal seat tracks and floor panels. Local variations in this floor structure include the area over the center wing box, and the main landing gear wheel well across which the floor beams run longitudinally, and the control cabin where the floor structure has to accommodate control gear and other special equipment. When installed, the forward air stairs are contained below the floor of body section 41. The fuselage frames at body stations 540 and 664 incorporate points at which the fuselage is attached to the wing front and rear spars. The connection between the inboard end of the landing gear support beam and the fuselage is a swinging link fitting attached to the frames at body stations 695 and 706.

C. Body section 48 of the fuselage is not pressurized and extends aft from the rear pressure bulkhead at body station 1016. The vertical fin structure and the horizontal stabilizer structure are supported by the 48 section. The APU is installed in a fire proof compartment below the horizontal stabilizer (Fig. 4-7).

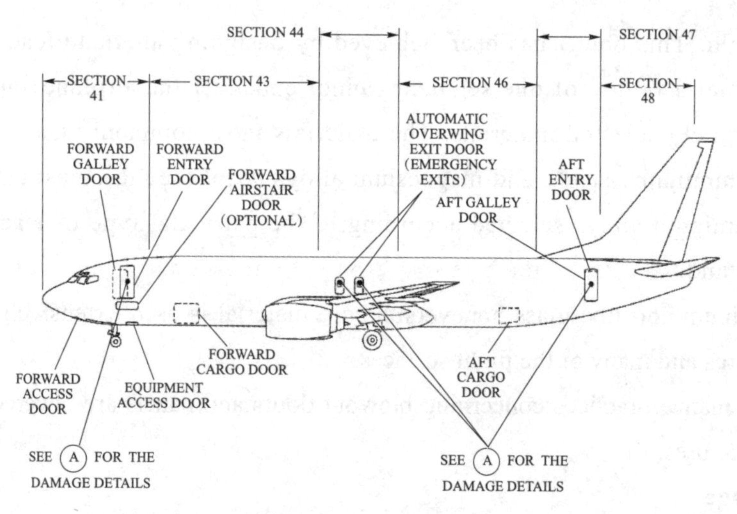

Fig. 4-7 Fuselage

Selected from AMM

Words for Reference

circumferential	[sə,kʌmfə'renʃəl]	*adj.* 沿边缘的；（尤指）圆周的
longitudinal	[,lɒndʒɪ'tjuːdɪnl]	*adj.* 纵向的；经度的
stringer	['strɪŋə]	*n.* 纵梁；桁条
empennage	['empɪnɪdʒ]	*n.* 尾部；尾翼
semi-monocoque	['semɪ'mɒnəʊkɒk]	*n.* 半硬壳式
aluminum	[ə'luːmɪnəm]	*n.* 铝
magnesium	[mæg'niːzɪəm]	*n.* 镁（化学元素）
surmount	[sə'maʊnt]	*vt.* 克服；越过
seat track		座椅轨道
spar	[spaː]	*n.* 翼梁
bulkhead	['bʌlk,hed]	*n.* 隔板；舱壁

Reading Tasks

1. Judge the following statements according to the contents of the AMM.

(1) Aluminum alloys are the most extensively used in manufacturing airplane. ( )

(2) Our primary concern in selecting aviation materials is the strength and weight of the material. ( )

(3) The APU is installed in a fire proof compartment below the vertical stabilizer. ( )

2. Fill in the blanks according to the above reading materials.

(1) Aluminum and fiberglass honeycomb core material is used extensively on _______ areas of structures and many of the flight surfaces.

(2) The fuselage is a semi-monocoque structure with the _______ reinforced by

circumferential ________ and transverse _________.

Translate the following sentences into Chinese.

1. The structure of the airplane is designed to provide maximum strength with minimum weight.

____

2. The fourth section of the fuselage is at the aft end and provides support for the empennage.

____

3. The vertical fin structure and the horizontal stabilizer structure are supported by the 48 section.

____

4. The fuselage frames at body stations 540 and 664 incorporate points at which the fuselage is attached to the wing front and rear spars.

____

5. When installed, the forward air stairs are contained below the floor of body section 41.

____

Translate the following abbreviations into corresponding Chinese.

1. STA____________
2. MLG____________
3. Max____________
4. Mini____________
5. Al____________

Part III Tools and Equipments

Files

1. Match the words or phrases with the translations.

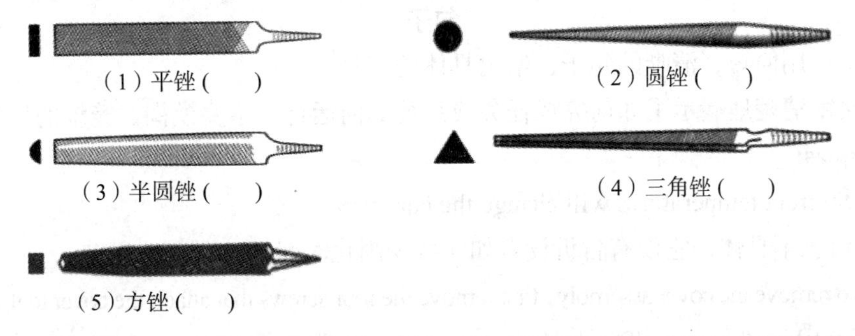

A. Square file
B. Flat file
C. Triangular file
D. Round file
E. Half-round file

2. **Fill in the words or phrases in the blanks.**

Square file	Triangular file	Round file
Half-round file	Flat file	

______ (1) This file is of rectangular cross-section in shape. It is used to smooth flat surfaces or large curves by filing and finishing the workpiece.

______ (2) This type of file has a round section. It is used for the purpose of rubbing or finishing the holes of a small diameter.

______ (3) This file has one flat side and the other side is the shape of a semicircle (rounded). It is mostly used for smoothening, stock removal, filing internal curved surfaces, and finish surfaces. Its half-round shape allows deburring inside curved spaces, round holes, and corners.

______ (4) This is a type of file that has a triangular cross section. The file has three tapered sides, which are used to make sharp corners that are smooth and flat.

______ (5) This type of file is in the shape of a square. Filing in a rectangular and square groove can be done by using it.

Part IV Simplified Technical English

Sentences

句子

★规则：用简短、清晰的句子，给出具体的信息。

确保文字清楚地展示了如何完成任务或系统如何运行，不要模糊、笼统的信息。

Examples:

(1) Different temperatures will change the cure time.

(句子信息不具体，它没有告诉读者如何减少固化时间。)

(2) To remove the cover assembly, first remove the four screws that attach the cover to the housing, and then, after taking the cover off the housing, remove the preformed packing and throw it away.

在程序性写作中，直接针对读者，给出简短、清晰的说明。可将上一段描述改写如下：

Remove the cover assembly as follows:

1. Remove the four screws that attach the cover to the housing.

2. Remove the cover from the housing.

3. Remove and discard the preformed packing.

Vertical Lists
垂直列表

★规则：当你必须在一个句子中包含许多不同的项目或动作时，为了让句子更容易阅读和理解，可使用垂直列表。

注意事项：

垂直列表中的每一项都用大写字母开头。完整的句子，在垂直列表中末尾加上句号。不是完整的句子，在垂直列表的最后一项末尾加一个句号。

Examples:

The wheel assembly comprises the tire, the tube, the spoke, the spoke fittings, the valve, and the hub.

The wheel assembly has these parts:

-The tire

-The tube

-The spokes

-The spoke fittings

- The valve

- The hub

Part V Maintenance English Application Practice

Choose the best answer from the following choices.

( ) 1. The control unit is installed aft of the cockpit.

A. 该控制组件安装在机身尾部。

B. 该控制组件安装在驾驶舱后部。

C. 该控制组件位置在驾驶舱底部。

( ) 2. If the wheel temperature increases to more than 177 degree, the core of the fusible plug melts and the tire deflates.

A. 如果轮胎温度升高超过 177 摄氏度，插头熔化且轮胎放气。

B. 如果轮胎温度升高超过 177 摄氏度，熔塞和轮胎易融化。

C. 如果轮胎温度升高超过 177 摄氏度，熔塞易熔化且轮胎放气。

( ) 3. Equipment without significant effect on the weight or balance of the airplane need not be listed.

A. 设备对没有列出的飞机的质量或平衡没有显著影响。

B. 对飞机质量或平衡没有显著影响的设备不需要列出。

C．没有明显影响的设备的质量或平衡在没有列出的飞机上。

（　）4．The airframe manufacturer will incorporate into the maintenance manual, where appropriate, the "shared interest" data supplied by component manufacturers.

A．机身制造商将把部件制造商提供的合适的“共同利益”资料纳入维护手册。

B．在适当情况下，机身制造商将把部件制造商提供的维护手册与“共同利益”资料进行合并。

C．机身制造商将在适当情况下把部件制造商提供的“共同利益”资料纳入维护手册。

（　）5．Tire wear increases with high speed.

A．安装的轮胎随速度的增加而增加。

B．轮胎的磨损将会增加速度。

C．轮胎磨损随速度的增加而增加。

（　）6．这些测试应不需要安装在飞机上以外的任何特殊设备或设施，并应与机组执行的测试相比较。

A．These tests should require no special equipment or facilities rather than that installed on the aircraft and should be comparable to the test performed by the flight crews.

B．These tests should require no special equipment or facilities other than that installed on the aircraft and should be comparable to the test performed by the flight crews.

C．These tests should require none special equipment or facilities other than that installed on the aircraft and should be comparable to the test performed by the flight crews.

（　）7．当进行每一步的活门测试时，确保没有泄漏、没有腐蚀、没有生锈并且在整个程序结束后再次确认。

A．While you do each step of the valve test, make sure that there is no leakage, no corrosion, no rust and do the double check AFT the whole procedure again.

B．While you do one step of the piston test, make sure that there is no leak, no corruption, no rust and do the double check through the whole procedure again.

C．While you do each step of the valve test, ensure that there is no leakage, no corruption, no rust and do the double check AFT the whole procedure again.

（　）8．系统内的所有流体控制装置，如单向活门、燃油泵、储压器和继电器均应包括在内。

A．All flow control devices within the system, such as check valves, fuel pumps, accumulators, and relays should be included.

B．All flue control devices within the system, such as single valves, fuel pumps, storage, and relays should be contained.

C．All fluid control devices within the system, such as single valves, fuel pumps, accumulators, and relays should be involved.

（　）9．外界大气温度是指从机上温度指示或地面气象来源获得的自由空气静态温度，并针对仪表误差和可压缩性影响进行了调整。

A. Ambient air temperature is the free static air temperature, attained from both inflight temperature indications and surface meteorological sources, adjusted for instrument error and compressibility effects.

B. Outside air temperature is the free air electrostatic temperature, contained neither from inflight temperature indications nor ground meteorological sources, adjusted for instrument error and compressibility effects.

C. Outside air temperature is the free air static temperature, obtained either from inflight temperature indications or ground meteorological sources, adjusted for instrument error and compressibility effects.

(　) 10. Connect the towing arm to the nose gear.

A. Hinge the towing arm up to the nose gear.

B. Hook the towing arm up to the nose gear.

C. Fix the towing arm tip to the nose gear.

(　) 11. Those units provide a means of preventing it from freezing.

A. Those units means that they can prevent it from eroding.

B. Those units provide a method of deicing.

C. Those units provide a way to prevent the formation of ice.

(　) 12. These tests should require no special equipment or facilities other than that installed on the aircraft and should be comparable to the tests performed by the flight crews.

A. These tests should only use the equipment or facilities installed on the aircraft.These tests should be able to perform by the flight crews.

B. These tests should not use the special equipment or facilities that are not installed on the aircraft.These tests should be similar to the tests performed by the flight crews.

C. These tests should require the equipment or facilities installed on the aircraft rather than the special ones and should be replaceable by the tests accomplished by the flight crews.

(　) 13. Position 2 of the switch gives a signal of the aircraft airborne condition.

A. The symbol on the switch of position 2 indicate the airborne condition of the aircraft.

B. Switch number 2 in position shows the airborne condition of the aircraft.

C. Position 2 of the switch tells the aircraft airborne condition through a signal.

(　) 14. The doors inside the fuselage are installed in the fixed partitions. Include items such as structure, latching mechanisms, handles, lining, etc. Does not include doors installed in the movable partitions which are covered in Chapter 25.

A. Doors inside the fuselage, includes items such as structure, handles, lining, are installed in the fixed partition. Doors installed in the movable partitions which are covered in Chapter 25 are included as well.

B. The doors inside the fuselage installed in the fixed partitions involves items such as structure, latching mechanisms, handles, lining, etc. Doors installed in the movable partitions which are covered in Chapter 25 are excluded.

C. The doors inside the fuselage installed in the fixed partitions excludes items such as structure, latching mechanisms, handles, lining. Include doors installed in the movable partitions which are covered in chapter 25.

(   ) 15. Existing analyses were reviewed to identify component and areas that may have exhibited lower margins of safety.

A. To identify component and areas that may have exhibited lower margins of safety, in hand analyses were checked.

B. Analyses were checked to identify available component and areas where exhibited lower margins of safety may exists.

C. To identify in hand component and areas, analyses were checked where exhibited lower margins of safety may exists.

Part VI Safety and Regulations

China Aviation Hero Crew

On May 14, 2018, the Sichuan Airlines flight 3U8633 carrying 119 passengers departed from Southwest China's Chongqing municipality to Lhasa, capital of Tibet autonomous region. It was forced to make an emergency landing after a cockpit window broke at 9, 800 meters, with the co-pilot sucked halfway out of the aircraft. All passengers are safe, although the co-pilot sustained injuries to the face and waist, and another crew member was slightly hurt during the emergency landing, according to the Civil Aviation Administration of China (CAAC). All the flight crew were named as a "China Aviation Hero Crew" and the pilot, Liu Chuanjian, was awarded the title of "hero captain of China's civil aviation" by Civil Aviation Administration of China and Sichuan provincial government.

On September 30, 2018, General Secretary of the CPC Central Committee Xi Jinping met with nine Sichuan Airlines crew members of flight 3U8633. He said safety is the lifeline of the civil aviation industry and should be given great attention at all times. Civil aviation authorities and service providers should correctly handle the dynamic between safety and profit.

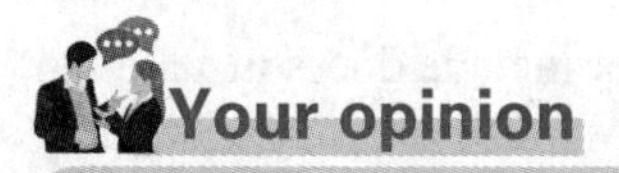

Without regular training, the heroic crew cannot complete a safe forced landing. While paying tribute to the heroic crew of Sichuan Airlines 3U8633, what should we learn from them?

Project 2　The Power Plant

Objectives

To grasp the basic knowledge of the power plant.

To understand the engine hazards of the power plant.

To master the usage of measuring tools.

To perceive the C919.

Part I　Aviation Situational Communication

1. Read aloud the topic-related sentences for aviation communication and try to match them with their Chinese meaning.

Listen

_______ (1) There is an engine start problem.

_______ (2) The left engine start valve can not open.

_______ (3) The start procedure is completed.

_______ (4) We are ready for the engine start.

_______ (5) We'll wait for your signals.

A. 启动程序已完成。

B. 左发启动活门无法打开。

C. 我们已准备好启动发动机。

D. 我们将等待您的信号。

E. 发动机启动有问题。

2. Listen to some short dialogues between the Pilot and the Ground mechanic. After the dialogues there are some questions. Please choose the most appropriate answer from the choices.

(1) Situation one

Listen

1. What problem did CNS123 encounter? (　　)

A. Engine fuel.　　C. Engine oil.

B. Engine start.　　D. Engine flameout.

2. What component was abnormal? (　　)

A. Engine starter.　　C. Left engine start valve.

B. Right engine start valve.　　D. EEC.

3. What does the pilot need to do? (　　)

A. Start the engine electrically.　　C. Start the engine automatically.

B. Start the engine mechanically.　　D. Start the engine manually.

(2) Situation two

1. What do the ground mechanics have to get? (　　)

A. Flash light.　　C. Test equipment.

B. Some tools.　　D. Protection equipment.

2. What do the ground mechanics wait for? (　　)

A. Signals to open the start valve.　　C. Signals to release.

B. Signals to shutdown the engine.　　D. Signals to open the shutoff valve.

Part II　Aviation Reading

2.1　Aviation Reading

Pre-reading Questions

1. What is the function of a power plant?

____

2. What are included in a power plant?

____

3. What are the main categories of engines?

____

The Power Plant

An airplane can fly because it has a forward motion of thrust, resulting in the relative movement of the airplane against the airflow, which is the result of lift. Therefore, the power plan of the aircraft is the core part of the aircraft (Fig. 4-8).

The Power Plant

Fig. 4-8 Aircraft Engine

The aviation power plant includes the aircraft engine and the necessary systems and accessories to ensure its normal operation. Its composition depends on the type of engine used, mainly the following systems or devices: Aircraft engine and its starting and control system; aircraft fuel system; aircraft lubrication oil system; fire prevention and suppression systems; aircraft engine cooling device; aircraft engine fixtures; intake and exhaust devices.

Aero-engines fall into two categories: Piston engines and jet engines (Fig. 4-9). Aviation piston engine is mainly four stroke gasoline internal combustion engine, light weight, high power. It was not until the advent of the jet engine in the late 1930s that aircraft power plants took on a different form. Since the first successful test flight of an aircraft equipped with a turbojet in Germany on September, 1939, the aviation gas turbine engine has developed rapidly. Gas turbine engine has simple structure, light weight, large thrust and high propulsion efficiency. In a wide range of flight speed, engine thrust increases with the increase of flight speed.

Fig. 4-9 Piston engines vs. jet engines

Most modern high-speed aircraft use jet engines. Only in small, low speed aircraft, piston engine is still in extensive use because of economic performance and easy maintenance. Turbofan engines are most commonly used in medium and large passenger and transport aircraft. Turboprop engines are also widely used in small and medium-sized subsonic aircraft.

The most commonly used engines in modern aircraft are turbojets and turbofans.

A turbojet engine consists of an inlet, a compressor, a combustion chamber, a turbine, and a nozzie (Fig. 4-10). With the characteristics of light weight, small size and large power, it is suitable for supersonic flight. When the engine is working, the air is compressed by the compressor, the pressure rises, and then enters the combustor to mix with the fuel and burn. The gas formed after combustion flows into the turbine, which then rotates under the drive of high temperature and high pressure gas, thus driving the compressor to work. The gas finally expands and accelerates in the nozzle, and ejects outwards at high speed to produce thrust. Fuel is added in the nozzle behind the turbine of the engine to form an afterburner, which can greatly improve the thrust, but in the range below high subsonic speed, the propulsion efficiency is low, and the fuel consumption increases more. The characteristic of turbojet engine makes it suitable for supersonic fighter jet power plant.

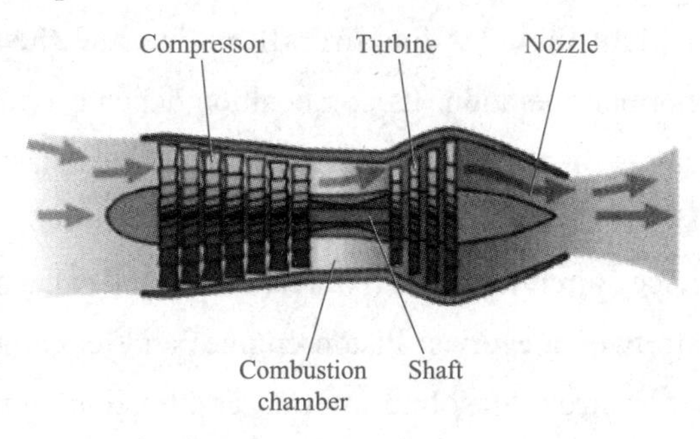

Fig. 4-10 Turbojet engine

The turbofan engine adds a fan channel (or outer culvert) outside the turbojet engine. The airflow in the fan channel has a low temperature and low speed. It mixes with the gas jet in the internal channel to reduce the exhaust velocity of the whole engine, greatly improving the propulsion efficiency of the engine in the high subsonic range, and reducing the engine noise at the same time. It has the characteristics of "high supercharging ratio, high turbine front temperature and high bypass ratio" (Fig. 4-11).

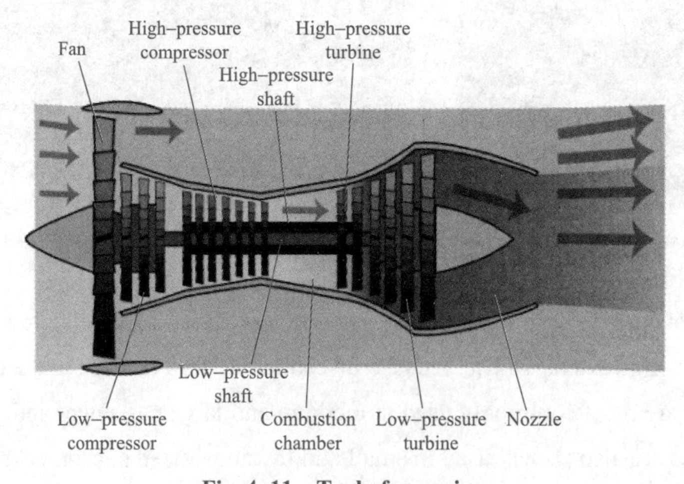

Fig. 4-11 Turbofan engine

Words and Expressions

thrust	[θrʌst]	*n.* 驱动力，推力
airflow	[ˈeəfləʊ]	*n.* 气流（尤指飞机等产生的）；空气的流动
intake	[ˈɪnteɪk]	*n.* 摄入，吸入；入口，进口
exhaust	[ɪgˈzɔːst]	*n.* 排气，排气装置
combustion	[kəmˈbʌstʃn]	*n.* 燃烧
turbine	[ˈtɜːbaɪn]	*n.* 涡轮机，汽轮机
turbojet	[ˈtɜːbəʊdʒet]	*n.* 涡轮喷气飞机
turbofan	[ˈtɜːbəʊˌfæn]	*n.* 涡轮风扇发动机
inlet	[ˈɪnlet]	*n.*（空气、气体或液体进入机器等的）进口，入口
compressor	[kəmˈpresə(r)]	*n.* 压缩机
chamber	[ˈtʃeɪmbə(r)]	*n.* 房间，室；腔，膛
nozzle	[ˈnɒzl]	*n.* 管口，喷嘴
supersonic	[ˌsuːpəˈsɒnɪk]	*adj.* 超声速的 *n.* 超声速，超声波；超声速飞机
accelerate	[əkˈseləreɪt]	*v.*（使）加快，促进；加速
consumption	[kənˈsʌmpʃ(ə)n]	*n.* 消费，消耗；食用，引用，吸入；专用
channel	[ˈtʃæn(ə)l]	*n.* 频道，电视频道；管道，通道，航道
velocity	[vəˈlɒsəti]	*n.* 速度，速率；高速，快速
bypass ratio		涵道比

Fill in the blanks with the proper words given below, changing the form if necessary.

thrust	combustion	turbine	compressor	velocity

1. It provides the ________ that makes the craft move forward.
2. The ________ turns a generator.
3. This air ________ doesn't work.
4. Acceleration and ________ are both vectors.
5. Invented in 1951, the rotary engine is a revolutionary concept in internal ________.

2.2 Manual Reading

There are some tasks for you to fulfill. You should read the Manual Reading materials carefully and then finish the tasks.

POWER PLANT—ENGINE HAZARDS

General

It is dangerous to work around engines. Use the entry/exit corridor when the engine is in operation. Also, stay out of the inlet and exhaust areas when the engine is in operation.

CAUTION: PERFORM FOR WALK IN FRONT OF AND AROUND ENGINE INGESTION AREA PRIOR TO ENGINE START.

These are the hazards around an engine in operation:

- Inlet suction
- Exhaust heat
- Exhaust velocity
- Engine noise

Inlet Suction

Engine inlet suction can pull people and large objects into the engine. At idle power, the inlet hazard area is a 10 ft (3.1 m) radius around the inlet (Fig. 4-12).

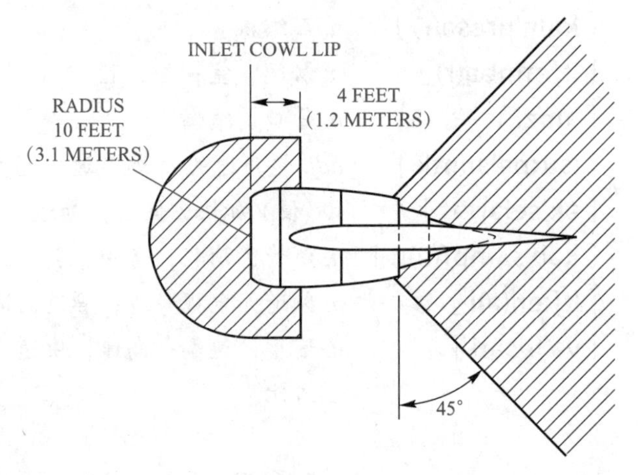

Fig. 4-12 Engine hazard at idle power

WARNING: IF THE WIND IS OVER 25 KNOTS, INCREASE THE INLET HAZARD AREA BY 20 PERCENT.

Exhaust Heat

The engine exhaust is very hot for long distances behind the engine. This can cause damage to personnel and equipment.

Exhaust Velocity

Exhaust velocity is very high for long distances behind the engine. This can cause damage to personnel and equipment.

Engine Noise

Engine noise can cause temporary and permanent loss of hearing. You must wear ear protection when near an engine in operation.

Engine Entry/Exit Corridor

Engine entry corridors are between the inlet hazard areas and the exhaust hazard areas. You should go near an engine in operation only when:

- Engine is at idle.

• You can speak with people in the flight compartment.
• For additional safety, wear a safety harness when the engine is in operation.
• Training Information Point.
• The beacon light must be on while the engines are on.

Selected from AMM 71

Words for Reference

hazard	['hæzəd]	*n.* 危险
corridor	['kɒrɪdɔː(r)]	*n.* 通道
exhaust	[ɪg'zɔːst]	*n.* 废气
velocity	[və'lɒsəti]	*n.* 速度
radius	['reɪdiəs]	*n.* 半径
beacon	['biːkən]	*n.* 信标
safety harness		安全带

Reading Tasks

1. Judge the following statements according to the contents of the AMM.

(1) Engine entry corridors are between the inlet hazard areas and the exhaust hazard areas. ( )

(2) The engine exhaust can cause damage to personnel and equipment. ( )

(3) You can go near an engine in operation when engine is not at idle. ( )

2. Fill in the blanks according to the above reading materials.

(1) Use the entry/exit ________ when the engine is in operation.

(2) The engine exhaust is very ______ for long distances behind the engine.

(3) Engine noise can cause temporary and ________ loss of hearing.

Translate the following sentences into Chinese.

1. Engine inlet suction can pull people and large objects into the engine.

____

2. If the wind is over 25 knots, increase the inlet hazard area by 20 percent.

____

3. You must wear ear protection when near an engine in operation.

____

4. For additional safety, wear a safety harness when the engine is in operation.

____

5. The beacon light must be on while the engines are on.

____

Part III Tools and Equipments

Common Measuring Tools

1. Match the words or phrases with the translations.

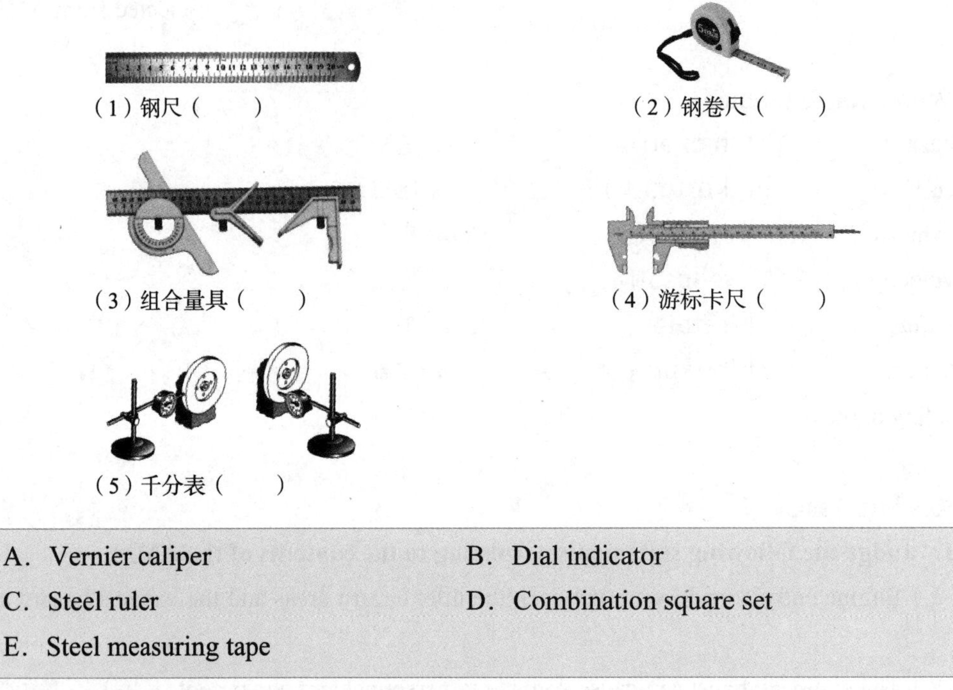

(1)钢尺(　　)

(2)钢卷尺(　　)

(3)组合量具(　　)

(4)游标卡尺(　　)

(5)千分表(　　)

A. Vernier caliper	B. Dial indicator
C. Steel ruler	D. Combination square set
E. Steel measuring tape	

2. Fill in the words or phrases in the blanks.

Dial indicator	Steel measuring tape
Vernier caliper	Steel ruler
Combination square set	

(1) It is a simple measuring instrument that is used for measuring distances and ruling straight lines. It is used to measure the length and size of parts, but its measurement results are not very accurate.

(2) This measuring instrument consists of a steel ruler and three moving heads called square head, protractor head and center head. The three moving heads can slide along the steel ruler and be clamped at any desired location. By removing all the heads, the steel ruler may be used alone as a rule or a straight edge. It is a very important instrument may be used for a great many purposes, such as measuring length, depth, height, inclination, levelness, angle, and center of circle.

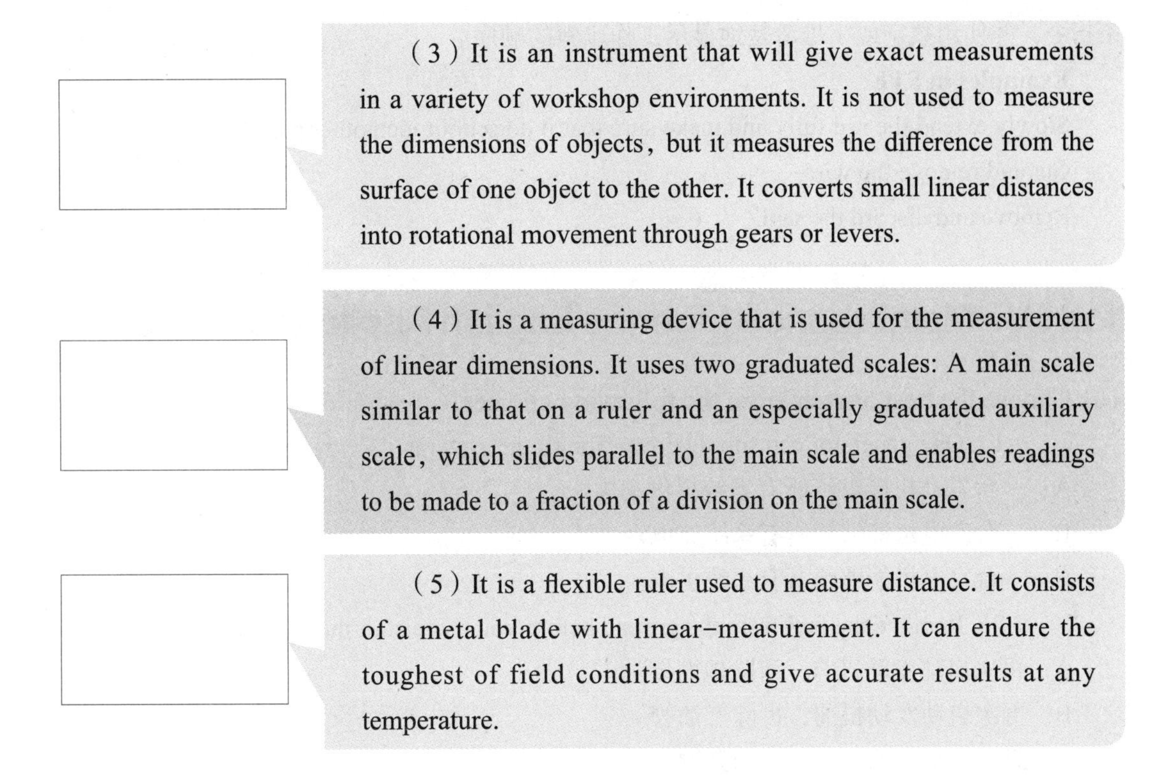
(3) It is an instrument that will give exact measurements in a variety of workshop environments. It is not used to measure the dimensions of objects, but it measures the difference from the surface of one object to the other. It converts small linear distances into rotational movement through gears or levers.

(4) It is a measuring device that is used for the measurement of linear dimensions. It uses two graduated scales: A main scale similar to that on a ruler and an especially graduated auxiliary scale, which slides parallel to the main scale and enables readings to be made to a fraction of a division on the main scale.

(5) It is a flexible ruler used to measure distance. It consists of a metal blade with linear-measurement. It can endure the toughest of field conditions and give accurate results at any temperature.

Part IV Simplified Technical English

Procedural writing
程序性写作

★规则：尽可能地保持句子的简短性，最长不要超过 20 个单词。

Examples:

Non-STE: Put preservation oil into the unit through the vent hole until the oil level is approximately 6 mm (0.24 in.) below the surface of the flange cover.

STE: Put preservation oil into the unit through the vent hole. Continue until the oil level is approximately 6 mm (0.24 in.) below the surface of the flange cover.

第一句改为第二句的差异性是什么?

第一个句子现在分成了两个较短的句子，读起来比较容易理解。修改的句子中，第一个句子有 10 个字，第二个句子有 16 个字。该指令不能分为两个工作步骤，因为必须同时执行这两个操作。

★规则：除非两个或多个动作同时发生，否则每个句子中只写一个动作。

如果一个句子中有太多的指令，这个句子就不容易理解。在每个工作步骤中只设置一条指令，并使用数字或字母清楚地显示工作步骤的顺序。

Examples in STE:

Slowly extend the rod fully and make sure that it does not touch other parts.

Cut and remove the wire.

Remove and discard the seal.

Part V Maintenance Licence English Application Practice

Choose the best answer from the following choices.

( ) 1．One generator can supply power for all the systems.

A．一台发动机能够给所有系统提供动力。

B．一台发电机能够给所有系统充电。

C．一台发电机能够给所有系统供电。

( ) 2．Remove the fuel from the ground before you try to start the engine again.

A．再次启动发动机前，清除地面燃油。

B．再次启动发动机前，拆除燃油管。

C．启动发动机前，清除地面杂物。

( ) 3．Motor the engine without ignition.

A．发动机的驱动马达没有点火。

B．启动发动机但不点火。

C．带转发动机但不点火。

( ) 4．If the airplane has a built-in auxiliary power unit，describe it and explain how it fits in with the electrical system.

A．如果飞机有一个内置的辅助动力装置，请对其进行描述并说明其如何与电气系统配合使用。

B．如果飞机有一个附加的辅助电源组件，描述该装置并解释它是如何对接电气系统的。

C．如果飞机有一个内置的辅助动力装置，描述该装置并解释它是如何装配到电气系统的。

( ) 5．Write the temperature on the engine record card.

A．在发动机记录卡上填写记录。

B．在室温下填写发动机记录卡。

C．把温度写在发动机记录卡上。

( ) 6．Accelerate-Stop Distance is the distance required to accelerate an airplane to a specified speed and，assuming failure of an engine at the instant that speed is attained，to bring the airplane to a stop.

A．中断起飞距离是使飞机加速至指定速度，并假设达到该速度的瞬间发动机发生故障，使飞机停下来所需的距离。

B．中断起飞距离是使飞机加速至特定速度，并假设达到该速度时发动机发生故障，使飞机停下来所需的距离。

C．中断起飞距离是使飞机加速至特殊速度，并即将达到该速度的瞬间发动机发生故障，使飞机停下来所需的制动距离。

（　）7．Procedures shall be provided for start the engine in flight and，in the event the engine does not start，for subsequent action（s）.

A．提供在飞行中启动发动机的程序，如果发动机不能启动，则应采取后续措施。

B．提供在飞行中启动发动机的程序，如果发动机能启动，则应采取后续措施。

C．在飞行中启动发动机及发动机不能启动的程序均应提供，则应采取补救措施。

（　）8．The exhaust gas temperature indicator，on piston engine powered airplanes，is the instrument used to identify the lean fuel flow mixtures for various power settings.

A．在活塞发动机驱动的飞机上，排气温度指示器用于识别各种功率设置下的油气比。

B．在活塞发动机驱动的飞机上，排气温度指示器用于设置各种功率设置下的油和空气的混合比。

C．在动力活塞驱动的飞机发动机上，排气温度指针用于识别最大功率设置下的油气比。

（　）9．若觉得发动机过热，将电门设置在“关”位置。

A．If the engine appears to be heating，turn it off.

B．If the engine appears to be overheating，turn it off.

C．If you think the engine is too hot，turn the switch to off.

（　）10．使用孔探仪从内部检查燃烧室。

A．Use a hole instrument to examine the turbine internally.

B．Use a borescope to examine the combustor internally.

C．Use a hole instrument to examine the combustor externally.

（　）11．动力装置仪表标记的说明应紧跟着动力装置限制的说明。

A．The sequence of the explanation should be powerplant instrument markings，then powerplant limitations.

B．An explanation of powerplant instrument markings shall immediately follow presentation on powerplant limitations.

C．Presentation on powerplant limitations shall come after the explanation of powerplant instrument markings.

（　）12．确保发动机进气道无外来物。

A．Ensure that there are no loose component in the engine air intakes.

B．Ensure that there are no loose dusts in the engine air intakes.

C．Make sure that there are no foreign objects in the engine air intakes.

（　）13．在重着陆过程中，发动机可能发生冲击载荷。

A．Shock loading of an engine may arise from a repeat landing.

B．Shock loading of the engine can occur during a light landing.

C. Shock loading of the engine can occur during a heavy landing.

(   ) 14. If it is possible that the temperature values will be more than the permitted limits, decrease the power.

A. If the temperature values are under the eligible limits, decrease the power.

B. Decrease the power as long as the temperature values are less than the certified limits.

C. If the temperature values are likely to be more than the allowed limits, decrease the power.

(   ) 15. Flight idle power is the power required to run an engine, in flight, at the lowest speed that will ensure satisfactory engine operation and airplane handling characteristics.

A. In flight, flight idle power required for the engine to operate at slowly speed is an important condition to ensure good engine operation and airplane handling characteristics.

B. Flight idle power is the power required for an engine to operate at a minimum speed during flight to ensure satisfactory engine operation and aircraft handling characteristics.

C. In flight, good engine operation and aircraft handling characteristics are guaranteed to provide the engine with flight idle power needed to operate at the lowest speed.

Part VI Safety and Regulations

The C919

The C919, full name COMAC C919, is a large civil jet developed by China in accordance with international civil aviation regulations and with independent intellectual property rights. It has a seat class of 158–192 and a range of 4, 075–5, 555 kilometers.

C919 development highlights:

On February 9, 2006, The State Council issued *The Outline of the National Program for Medium—and Long-Term Scientific and Technological Development* (2006–2020). The major project of large aircraft has been identified as one of the 16 major science and technology projects.

On February 26, 2007, The State Council held the 170th executive meeting, and in principle passed the *Large Aircraft Program Demonstration Report*, and in principle approved the major science and technology project of large aircraft formally.

On May 11, 2008, COMAC was established.

On January 6, 2009, Official release of the type code of the first single aisle conventional layout 150–seat class large passenger aircraft "COMAC919", short for "C919".

On December 9, 2011, the C919 large passenger jet project passed the national preliminary design review and entered the detailed design phase.

On November 2, 2015, the first C919 large passenger aircraft was officially rolled off the assembly line at Pudong Base, marking the phased achievement in the development of the C919 large passenger aircraft project.

The C919 makes its maiden flight in Shanghai on May 5, 2017.

On Sept 29, 2022, the C919 large passenger jet received the type qualification certificate issued by the Civil Aviation Administration of China.

The world's first large passenger jet, the C919, was delivered to China Eastern Airlines on Dec 9, 2022.

On May 28, the C919 made its first commercial flight from Shanghai Hongqiao Airport to Beijing Capital Airport. This historic event is of great significance to China's aviation industry and the overall economic development, and will form a far-reaching impact.

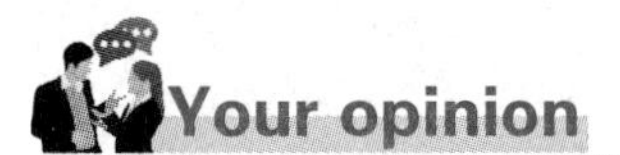

Your opinion

What significance does the C919 commercial debut have for China? What kind of spirit supports the successful completion of C919's commercial debut?

Project 3　Landing Gear

Objectives

To grasp the basic knowledge of the landing gear.

To understand the general description of the tricycle type landing gear.

To master the usage of gauges.

To perceive the sense of responsibility.

Part I　Aviation Situational Communication

1. Read aloud the topic-related sentences for aviation communication and try to match them with their Chinese meaning.

Listen

_______ (1) We have to circle.

_______ (2) We don't have a light on our nose gear yet.

_______ (3) Climb straight ahead.

_______ (4) The nose gear indicator light fails in the test.

_______ (5) The nose gear is not down and locked.

A. 前起落架指示灯在测试中失败。

B. 我们的前起落灯没亮。

C. 前起落架没有放下并锁定。

D. 我们必须盘旋。

E. 直接爬升。

Landing gear maintenance

2. Listen to some short dialogues between the Pilot and the Air Traffic Controller. After the dialogues there are some questions. Please choose the most appropriate answer from the choices.

(1) Situation one

1. Why does CNS123 have to circle? (   )

A. The nose gear indicator light is not on.

B. The main landing gear is up and locked.

C. The main landing gear is broken.

D. The nose landing gear is down and locked.

2. What does CNS123 need to do to know whether the bulb is probably burnt out? (   )

A. Check the landing gear.

C. Remove the bulb.

B. Replace the bulb.

D. Further test.

3. What is the suggestion of the Air Traffic Controller? (   )

A. Climb straight ahead to 200 feet.

C. Descend to 200 feet.

B. Climb straight ahead to 2, 000 feet.

D. Descend to 2, 000 feet.

(2) Situation two

1. What will the mechanic do? (   )

A. Check the nose gear.

C. Test the lights.

B. Ask for help.

D. Discuss the problem.

2. What are the possible results?

A. Both the light bulb and the landing gear could be broken.

B. The landing gear is not down and not locked.

C. The bulb is burned out.

D. The landing gear is down and locked.

Listen

Part II Aviation Reading

2.1 Aviation Reading

Pre-reading Questions

1. What's the purpose of the landing gear?

---

2. What does tricycle landing gear consist of ?

---

3. What are advantages of tricycle landing gear?

Landing Gear

The landing gear is also known as undercarriage. As one of the main parts of an airplane, the landing gear provides support for the airplane when it is on the ground, and allows the plane to take off, land, and taxi safely.

Tricycle is the most widely used landing gear configuration. Tricycle landing gear consists of two main landing gears and a nose landing gear. The wheels aft of the aircraft Center of Gravity (CG) is very close to it (compared with forward gear) and carry much of the aircraft weight and load, thus is referred to as the main wheel (Fig. 4-13). The forward gear, called the nose-gear, is far from CG (compared with main gear), hence it carries much smaller load (Fig. 4-14).

Landing Gear

Fig. 4-13 Main landing gears

Fig. 4-14 Nose landing gear

The main gear is located inboard of each engine, aft of the rear wing spar (Fig. 4-15), it provides the support for the aft section of the fuselage. The main gear absorbs impact on landing, and shocks and vibration while taxiing with a shock strut. Each main gear is hydraulically actuated to retract inboard into the fuselage.

Fig. 4–15　The location of main landing gear

As a steerable wheel assembly, the nose gear is located below the aft bulkhead of the flight compartment. The nose landing gear provides the support for the forward section of the fuselage as well as directional control on the ground. The nose landing gear usually uses a forward-retracting design, and it acts as a failsafe in the event of hydraulic pressure loss.

Airplanes with tricycle-type landing gear are directionally stable on the ground as well as during ground maneuver (turn). This arrangement enables the aircraft to have a fairly large crab angle during cross wind landing. The pilot view during take-off and landing is much better.

Commercial jet aircraft are frequently equipped with tricycle configuration.

Words and Expressions

undercarriage	[ˈʌndərˌkerɪdʒ]	*n.* 着陆装置
support	[səˈpɔrt]	*n.* & *v.* 支撑，支持
configuration	[kənˌfɪgəˈreɪʃ(ə)n]	*n.* 配置
load	[loʊd]	*n.* & *v.* 荷载，负担
failsafe	[ˈfeɪlˌseɪf]	*n.* 故障保护，失效保护
impact	[ɪmˈpækt]	*n.* & *v.* 冲击，撞击
taxi	[ˈtæksɪ]	*n.* & *v.* 滑行
jet	[dʒet]	*n.* 喷气式
tricycle landing gear		前三点式起落架
main landing gear		主起落架
nose landing gear		前起落架
wing spar		翼梁
shock strut		减震支柱

flight compartment 驾驶舱
crab angle 偏航角
CG *abbr.* 重心

Fill in the blanks with the proper words given below, changing the form if necessary.

support	impact	load	taxi	undercarriage

1. The C919 is ______ on the runway.
2. The main gear carry much of the aircraft weight and ____.
3. The landing gear ______ the airplane on the ground.
4. The ____ is one of the main parts of an airplane.
5. A shock strut can absorb____ when the aircraft is landing on the ground.

2.2 Manual Reading

There are some tasks for you to fulfill. You should read the Manual Reading materials carefully and then finish the tasks.

LANDING GEAR—GENERAL DESCRIPTION

General

The 737 airplane has a tricycle type landing gear with air/oil shock struts.

These are the landing gear structural systems:

- The main landing gear (MLG) and doors.
- The nose landing gear (NLG) and doors.

The landing gear extension and retraction systems extend and retract the main and nose landing gear.

The nose wheel steering system supplies the ground directional control of the airplane.

Components

The landing gear system has these main components:

- Control lever assembly
- Manual extension mechanism
- Transfer valve
- Selector valve
- Main landing gear (2)
- Nose landing gear
- Shimmy damper
- Proximity switch electronics unit (PSEU)
- Landing gear panel

• Auxiliary landing gear position lights

General Description

Hydraulic system A normally supplies pressure to the landing gear extension and retraction. Hydraulic system B supplies pressure for retraction only.

The landing gear transfer valve receives electrical signals from the proximity switch electronics unit (PSEU) . The landing gear transfer valve changes the pressure source of the landing gear from hydraulic system A to hydraulic system B.

You move the landing gear control lever assembly to control landing gear extension and retraction. The control lever moves the selector valve through cables.

The selector valve also gets an electrical input from the manual extension system. This operates a bypass valve in the selector valve to connect the landing gear retraction to the hydraulic system return. This lets the manual extension system extend the landing gear.

Landing gear lights show the position of the landing gear. The PSEU receives landing position signals from sensors on the landing gear. The normal and auxiliary lights are controlled by the PSEU.

The pressure for nose wheel steering comes from the nose landing gear extension pressure only. Hydraulic system A normally supplies pressure to the nose gear steering through the landing gear control system.

The landing gear control system also provides normal or alternate hydraulic pressure to these systems:

• Main landing gear shimmy damper.

• Gear retract brake system（Fig. 4-16）.

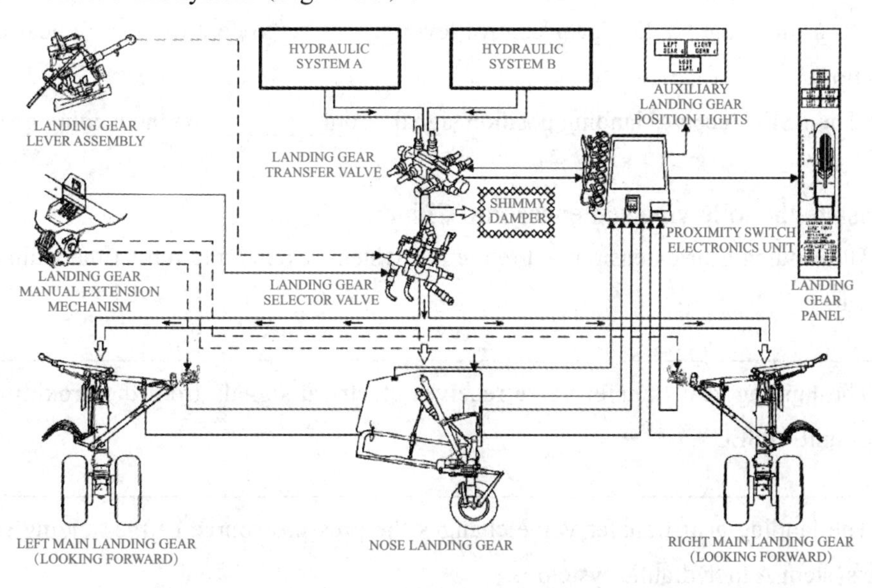

Fig. 4-16　General description of the landing gear system

Selected from AMM 32

Words for Reference

extension	[ɪk'stenʃn]	*n.* 延伸
retraction	[rɪ'trækʃn]	*n.* 收回
hydraulic	[haɪ'drɒlɪk]	*adj.* 液压的
cable	['keɪbl]	*n.* 电缆
sensor	['sensə(r)]	*n.* 传感器
shimmy damper		减摆器
bypass valve		旁通阀
MLG（main landing gear）		主起落架
NLG（nose landing gear）		前起落架
PSEU（proximity switch electronics unit）		接近电门电子组件

Reading Tasks

1. Judge the following statements according to the contents of the AMM.

(1) The ground directional control of the airplane is supplied by the nose wheel steering system. (　)

(2) Hydraulic system B normally supplies pressure to the landing gear extension and retraction. (　)

(3) The pressure for nose wheel steering comes from the nose landing gear and main landing gear extension pressure. (　)

2. Fill in the blanks according to the above reading materials.

(1) The 737 airplane has a ________ type landing gear with air/oil shock struts

(2) You move the landing gear control lever ________ to control landing gear extension and retraction.

(3) The PSEU receives landing position signals from ________ on the landing gear.

Translate the following sentences into Chinese.

1. The landing gear extension and retraction systems extend and retract the main and nose landing gear.

____

2. The landing gear transfer valve receives electrical signals from the proximity switch electronics unit (PSEU).

____

3. The landing gear transfer valve changes the pressure source of the landing gear from hydraulic system A to hydraulic system B.

____

4. The control lever moves the selector valve through cables.

____

5. Landing gear lights show the position of the landing gear.

____

Translate the following abbreviations into corresponding Chinese.

1. MLG ________________
2. NLG ________________
3. PSEU ________________

Part III Tools and Equipments

Common Measuring Tools

1. Match the words or phrases with the translations.

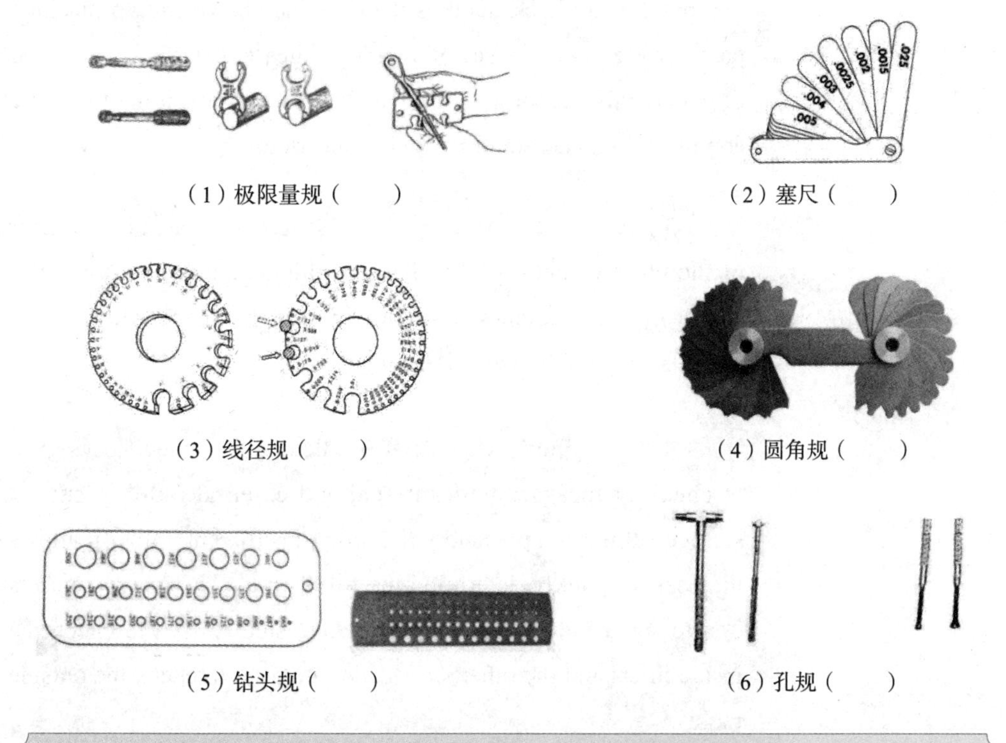

（1）极限量规（　　）　　（2）塞尺（　　）

（3）线径规（　　）　　（4）圆角规（　　）

（5）钻头规（　　）　　（6）孔规（　　）

A. Feeler Gauge	B. Bore Gauge
C. Limit Gauges	D. Wire Gauge
E. Fillet Gauge	F. Drill Bit Gauge

2. Fill in the words or phrases in the blanks.

Drill Bit Gauge	Limit Gauge
Wire Gauge	Bore Gauge
Feeler Gauges	Fillet Gauge

(1) It is an inspection tool without a scale, which is used to check the dimensions of manufactured components. It is different from a conventional measuring tool. As unable to use it to determine the actual size of dimensions of parts, we use a limit gauge to inspect whether a work piece is within the specified tolerance.

(2) They are mechanical measurement instruments that are used to provide a precise reading of the gap that exists between two parallel surfaces, such as the clearance between two machine parts or elements. A set of feeler gauges consist of a series of dimensionally accurate pieces of blades that are joined using a common shaft and nut or riveted connection.

(3) It is used to measure the cross-sectional area or diameter of the wire. Generally, the shape of this gauge is like a circular which includes notches through edges of the shape where every notch is stamped by wire size number.

(4) This gauge, which is also called radius gauge, is designed to check or measure both internal and external radii on curved surfaces. It is not a precision measuring instrument, rather it allows the user to quickly determine the fillet value. The gauge consists of two sets of blades on each side. One side is used to check the inside fillets and the other side blades are used to check the outside radius.

(5) It is a tool that measures the outside diameter of drill bits in different measurement systems, such as fractional, metric, English and letter. It can help to prepare a drill bit for accurate drill holes.

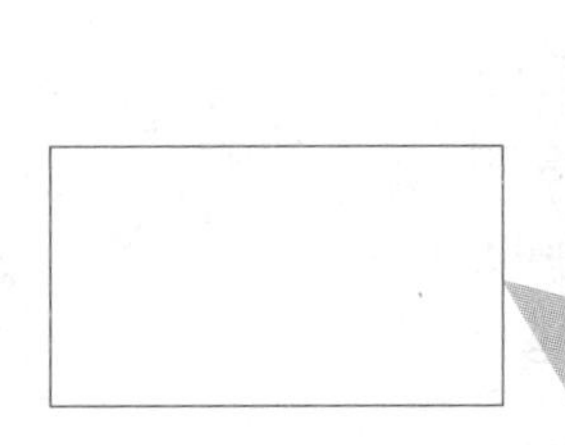

(6) It is a tool used to measure the inside of a bore, or hole. Once it is inserted into the hole that needs measuring, small parts called anvils expand outward to determine the diameter. It is a less precise, much simpler tool that requires measurement of a transferred dimension. There are two types: telescoping gauges and small hole gauges.

Part IV Simplified Technical English

Example procedural writing

程序性写作示例

★规则：用祈使语气写指令。

Examples in STE:

Set the switch to ON.

Remove the four bolts.

Increase the pressure to 60 psi.

Inflate the tires.

Install the new O-ring.

Examples:

Non-STE: The test can be continued.

STE: Continue the test.

Non-STE: Oil and grease are to be removed with a degreasing agent.

STE: Remove oil and grease with a degreasing agent.

Part V Maintenance English Application Practice

Choose the best answer from the following choices.

( ) 1. These parts can easily bend,break or become incorrectly aligned.

A. 这些部件不容易弯曲、断裂或对不齐。

B. 这些部件很容易掰断、断裂或对不齐。

C. 这些部件很容易弯曲、断裂或对不齐。

( ) 2. This installation procedure is typical for this type of fastener.

A. 此安装程序是此类紧固件的典型安装程序。

B. 此安装程序是此类松紧装置类型的安装程序。

C．此安装过程是此类紧固件的安装类型。

（　）3．如果由于腐蚀而无法拆卸螺栓，则使用渗透油。

A．If you cannot remove a nut because of corrosion,apply penetrating oil.

B．If you cannot disassembly a bolt because of corrosion,use penetrating oil.

C．If you cannot remove a bolt because of corrosion,apply penetrating oil.

（　）4．当进行活门测试的每一个步骤时，要确保没有渗漏。

A．While you do the valve test,ensure that there is no contamination.

B．While you do each step of the valve test,ensure that there is no blockage

C．While you do each step of the valve test,make sure that there is no leakage.

（　）5．磁性工具对罗盘系统有不利影响。

A．Magnetic tools have an adverse effect on the compass system.

B．Magnetic tools have an unwanted effect in the compass system.

C．Magnetic tools affect the compass system.

（　）6．这项工作适用于垂直安装。

A．This work is applicable to horizontal installations.

B．This work is applicable to direction installations.

C．This work is applicable to vertical installations.

（　）7．此程序仅适用于 A 型零件。

A．This procedure is only applicable to type A parts.

B．This task is only used for type A parts.

C．This procedure is applicable to all type parts.

（　）8．告知两个人准备操作灭火器。

A．Appoint two persons to be prepared to operate the fire extinguishers.

B．Tell two persons to be prepared to operate the fire extinguishers.

C．Talk two persons to be prepared to operate the fire extinguishers.

（　）9．将孔的直径扩至 8.00±0.003 毫米。

A．Machine the hole until it has a diameter of 8.00±0.003 mm.

B．Ream the hole until it has a diameter of 8.00±0.003 mm.

C．Spread the hole until it has a diameter of 8.00±0.003 mm.

（　）10．The first step in listing possible causes of each troubles to show what sections of the system may be at fault.

A．To show what sections of the system may be at fault,listing possible causes of each trouble 15 the first thing we should do.

B．What sections of the system may be at fault depends on the first step in listing possible causes of each trouble.

C．To show what sections of the system may be at fault is the first step when you list possible causes of each trouble.

Part VI Safety and Regulations

A Sense of Responsibility

On June 6, 1994, Northwest Airlines Flight WH2303 flew from Xi'an Xianyang International Airport to Guangzhou Baiyun International Airport. However, the plane disintegrated in the air at an altitude of 2, 884 meters, resulting in the death of 14 crew members and 146 passengers on board. The investigation results showed that the two plugs on the aircraft's autopilot were inserted backwards, so the heading damping and tilt damping could not work properly, and the plane could not maintain the flight attitude and rolled, resulting in stall and crash.

This is an air crash caused by an error of the aircraft maintenance personnel. If the maintenance personnel have a high sense of responsibility and carefully investigate the items, low-level errors such as the plug being inserted backwards are easily detected.

A good maintenance work style is the basis for safety. It is necessary to enhance the sense of responsibility and technical level of aircraft maintenance personnel, and reduce or even eliminate human errors in aircraft maintenance.

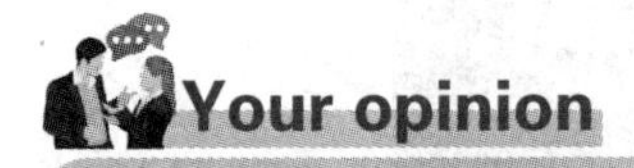

Your opinion

What is sense of responsibility and how should college students enhance their personal sense of responsibility?

Project 4　Aircraft Material

Objectives

To know the history of aviation materials.

To understand the general description of the material use in Boeing 787 and 777.

To master the usage of heating equipment.

To perceive the knowledge and identification of dangerous aviation materials.

Part I　Aviation Situational Communication

1. Read aloud the topic-related sentences for aviation communication and try to match them with their Chinese meaning.

________ (1) Something hit the aircraft.

________ (2) How serious is it?

________ (3) The aircraft was reported a major structural damage.

________ (4) We need an adviser to help the repair.

________ (5) We will contact the supplier to seek for help.

Listen

A. 我们需要一位顾问来协助修理。

B. 我们将联系供应商寻求帮助。

C. 有东西撞上了飞机。

D. 据报道，这架飞机有重大结构性损坏。

E. 有多严重？

2. Listen to some short dialogues between A and B. After the dialogues there are some questions. Please choose the most appropriate answer from the choices.

(1) Situation one

Listen

1. What did the truck hit? (　　)

A. AFT fuselage.　　C. Tail cone.

B. FWD fuselage.　　D. horizontal stabilizer.

2. What did the result of the accident? (　　)

A. A major structural damage.　　C. Damage to the right wing.

B. Leakage of fuel.　　D. Damage to the cargo door.

3. Who do we need to support the work? (　　)

A. Flight crew.　　C. mechanic.

B. Flight attendants.　　D. adviser.

(2) Situation two

1. When is the aircraft expected back into service? (　　)

A. by 18-July-2022.　　C. by 18-July-2020.

B. by 18-June-2022.　　D. by 18-June-2020.

2. How long will it take to finish the repair? (　　)

A. Not shown.　　C. 5 days.

B. 15 days.　　D. Based on the situation.

Part II Aviation Reading

2.1 Aviation Reading

Pre-reading Questions

1. When and who created the first powered aircraft in human history?

___

2. When did light aluminum monocoque structure come into use as load bearing structure?

___

3. What kind of materials are used in today's airplane structure according to the passage?

___

4. What Smart Materials does the article mention?

___

The History of Aviation Materials

"A generation material, a generation aircraft", to some extent, the development of aviation material determines the development of aviation industry. Actually, the materials used to make aircrafts are constantly being developed and improved. But the critical requirements for the materials are light and strong.

Early Airplane Structures

In 1903, the Wright Brothers succeeded in creating the first powered aircraft in human history (Fig. 4-17). It was made of wood frame, fabric and wire bracing. Wood frame accounts for 47% of the whole weight of aircraft. That's to say the main materials of the first airplane were wood and fabric.

The History of Aviation Materials

Fig. 4-17 The first powered aircraft

To make the plane stronger and more reliable, metal was used in airframe construction. In 1915, Junkers J-1 flew successfully. It was recognized as the first all-metal aircraft in the world. The main materials used to build the aircraft were iron and steel.

Because of the high density of steel, it is not practical to extensive use on aircraft. In order to obtain weight reduction, an aluminum alloy was developed. Aluminum alloy is lighter and more corrosion-resistant, which was the most suitable materials for aviation at that time. The chief limitation of it is low melting temperature. Light aluminum monocoque structure came into use as load bearing structure in the early 1930 s. During 1956 to 1969, titanium alloys were developed to overcome the high temperature intolerance of aluminum and the heavy weight of steel. Because titanium alloys are stronger and harder than aluminum, they are widely used in many high-temperature component structures, such as landing gears and joints. By the 1950's the transition to the full metal airplane had been completed.

Today's Airplane Structures

From 1970 to the present, composite materials have become a new member in the family of aeronautical structural materials. The airplane structure utilizes advanced composite materials for their high strength-to-weight properties (Fig. 4-18). As much as 25% weight savings have been realized on airplane components by using composite materials. The composite materials on the airplane consist of graphite, fiberglass or aramid Kevlar fibers, woven into a fabric form and pre-impregnated with a partially cured resin. When combined with Nomex honeycomb core material and fully cured, a high strength, high stiffness, and low weight structure results. Wing leading and trailing edge panels, control surfaces and wing-to-body fairings are constructed in this way. Panel edge bands and control surface spar and rib chords are constructed from laminate materials with no core.

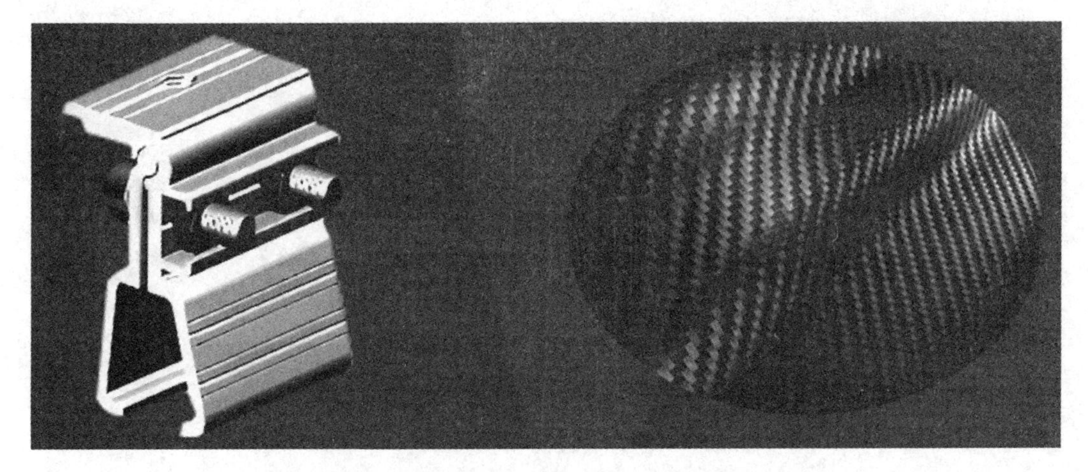

Fig. 4-18 Aluminum structure vs. composite structure

With the development of technology, many smart materials are emerging such as nanometer materials, self-healing materials, microlattice (Fig. 4-19). We are sure that more and more new materials will be used in aviation industry (Fig. 4-20).

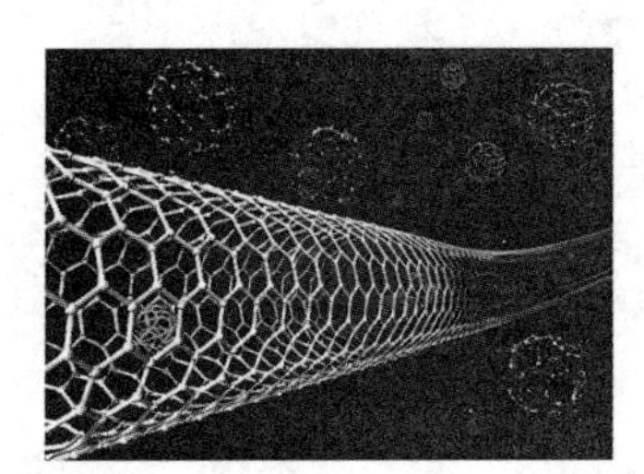

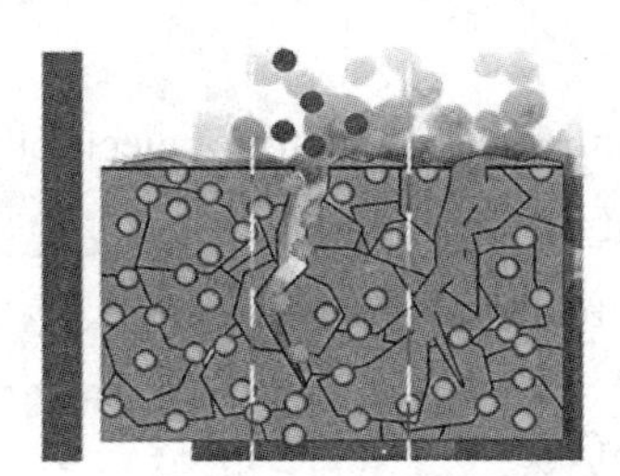

Fig. 4-19 Smart materials

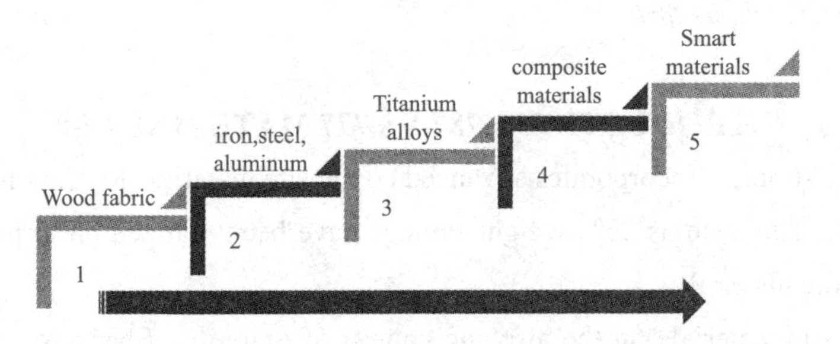

Fig. 4-20 The history of aeronautical materials

Words and Expressions

fabric	[ˈfæbrɪk]	*n.* 布料；织物 *adj.* 织物做成的；织物的
aluminum	[əˈluːmɪnəm]	*n.* 铝（化学元素）
alloy	[ˈælɔɪ]	*n.* 合金
titanium	[taɪˈteɪnɪəm]	*n.* 钛（化学元素）
graphite	[ˈgræfaɪt]	*n.* 石墨
fiberglass		玻璃纤维
Kevlar		凯夫拉
pre-impregnated		预浸渍的
resin	[ˈrezɪn]	*n.* 树脂
nanometer	[ˈnænəʊmiːtə]	*n.* 纳米
self-healing	[selfˈhiːlɪŋ]	*adj.* 自恢复性能的（自行净化的）
microlattice	[ˈmaɪkrəʊlætɪs]	*n.* 微格金属，微晶格
corrosion-resistant		耐腐蚀
load bearing structure		承重结构
strength-to-weight		强度质量比

Fill in the blanks with the proper words given below, changing the form if necessary.

fabric	graphite	alloy	fiberglass	resin

1. The window all changed into the aluminum ________.

2. The plastic ________ is used in a wide range of products, including electrical wire insulation.

3. ________ is a highly efficient conductor of electricity.

4. A composite rod is built with both graphite and ________ materials.

5. The ________ has been treated to repel water.

2.2 Manual Reading

There are some tasks for you to fulfill. You should read the Manual Reading materials carefully and then finish the tasks.

INTRODUCTION—787 VS 777 MATERIAL USE

The airplane structure incorporates advanced composite materials for their high strength-to-weight properties. As much as 25% weight savings have been realized on airplane components utilizing composite materials.

The composite materials on the airplane consist of graphite, fiberglass or aramid Kevlar fibers, woven into a fabric form and pre-impregnated with a partially cured resin. When combined with Nomex honeycomb core material and fully cured, a high strength, high stiffness, and low weight structure results. Wing leading and trailing edge panels, control surfaces and wing-to-body fairings are constructed in this way. Panel edge bands and control surface spar and rib chords are constructed from laminate materials with no core.

Before the 787, the 777 had the most composite construction of any model. The 777 utilized approximately 10 percent composite structure. The 787 uses over 50 percent composite structure. In addition to the overall increase in the use of composites on the 787, many composite component designs have changed from sandwich construction to solid laminate construction. Examples of this change in design are the flaps, flaperons and ailerons (Fig. 4-21).

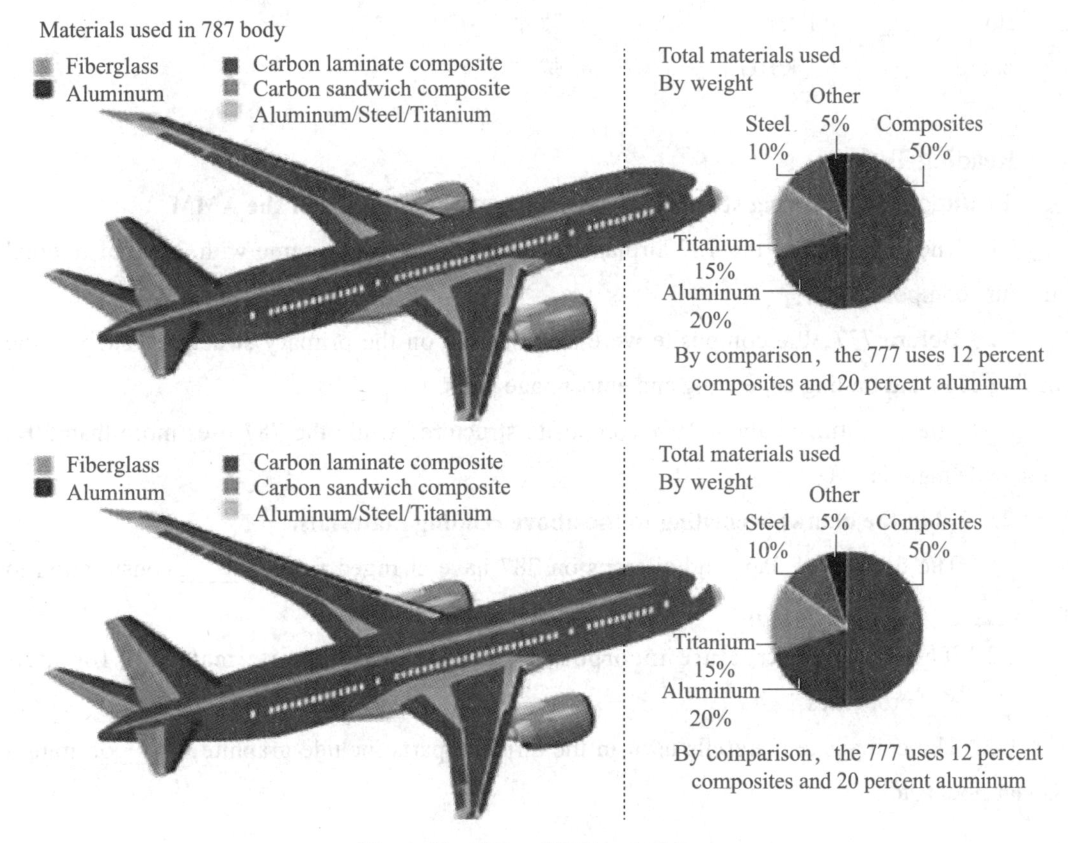

Fig. 4-21　787 and 777 material use

Prior to the 777, composites were primarily used on the leading edge and trailing edge wing and empennage parts. Composites were almost entirely used on secondary structure. Most of the composite structure on the 787 is primary structure.

Selected from 777 Training Manual

Words for Reference

VS (versus)　[ˈvɜːsəs]　*prep.* 对抗，与……相比

incorporate　[ɪnˈkɔːpəˌreɪt]　*vt.* 把……包括在内；将……纳入

utilize	[ˈjuːtəlaɪz]	*vt.* 使用；利用
graphite	[ˈɡræfaɪt]	*n.* 石墨
aramid	[ˈærəmɪd]	*n.* 芳族聚酰胺 （高强度阻燃合成纤维，用于制造消防服、防弹衣等）
pre-impregnated		预浸渍
Nomex		诺梅克斯（一种芳族聚酰胺纤维的商品名）
honeycomb core		蜂窝芯
laminate	[ˈlæmɪnɪt]	*n.* 层压板材
spar	[spɑː]	*n.* 圆材
rib	[rɪb]	*n.* 肋骨
chord	[kɔːd]	*n.* 弦

Reading Tasks

1. Judge the following statements according to the contents of the AMM.

(1) The materials on the 787 airplane exterior surface is the same with those of internal structure components. (   )

(2) Before 777, the composite were almost used on the primary structure, such as the leading edge and trailing edge wing and empennage parts. (   )

(3) The 777 utilized about 10% composite structure, while the 787 uses more than 50% composite materials. (   )

2. Fill in the blanks according to the above reading materials.

(1) The designs of flaps and ailerons on 787 have changed from ______ construction to solid _________ construction.

(2) The airplane structure incorporates advanced composite materials for their __________ properties.

(3) The composite materials used in the airplane parts include graphite, ____ or aramid Kevlar fibers etc.

Translate the following sentences into Chinese.

1. As much as 25% weight savings have been realized on airplane components utilizing composite materials.

____

2. The airplane structure incorporates advanced composite materials for their high strength-to-weight properties.

____

3. The 787 uses over 50 percent composite structure.

____

4. Most of the composite structure on the 787 is primary structure.

---

5. Many composite component designs have changed from sandwich construction to solid laminate construction.

Part III Tools and Equipments

Heating Equipment

1. Match the words or phrases with the translations.

（1）热压罐（　　）

（2）加热毯（　　）

（3）加热灯（　　）

（4）热气系统（　　）

A. Heat lamp	B. Heat blanket
C. Hot air system	D. Autoclave

2. Fill in the words or phrases in the blanks.

Heat lamp	Heat blanket
Hot air system	Autoclave

（1）It can be used to cure low temperature repairs, which cannot be used for high temperature repairs. Do not let the repair or the part touch it, or the pats would be damaged.

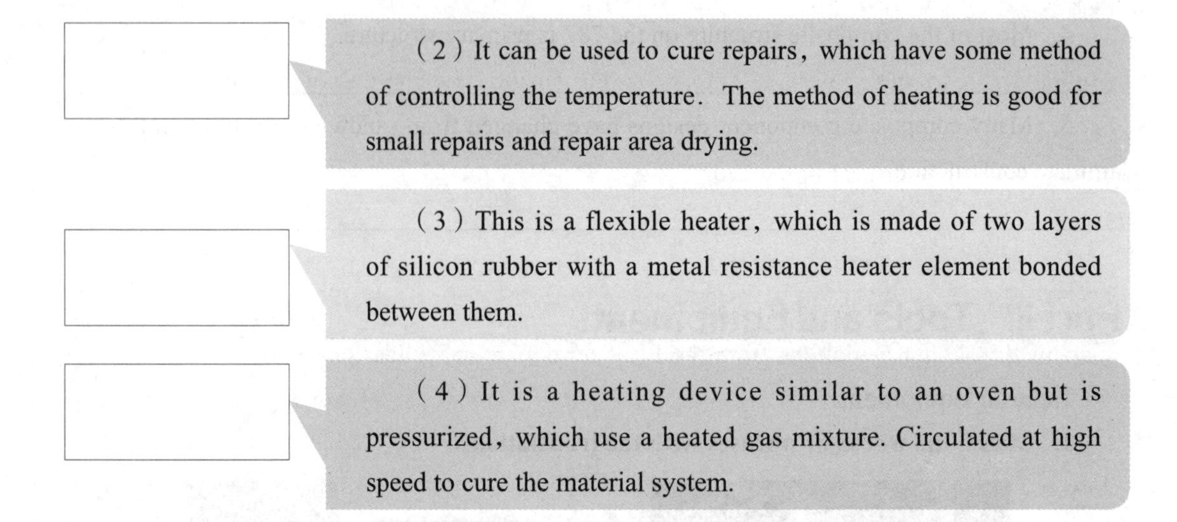

Part IV Simplified Technical English

Descriptive writing
描述性写作

描述性写作提供信息，而不是指导。

描述性写作可以是：

项目、产品、系统或组件的描述、功能、制造方法和操作方法。

提供一般信息的文本、程序中的注释。

★规则：循序渐进地提供信息。

在描述性文本中，逐步给出信息，并确保每个句子只包含一个主题。

Example:

First version（Non-STE）

Instrument Landing System

During the approach to the runway, deviation pointers in the course indicators give commands to fly up or down and left or right. This information comes from the VHF transceivers, which are part of the Instrument Landing System. This helps the pilot during the landing approach. When the pilot responds to the commands, the aircraft can be flown over the runway centerline (localizer) and at a fixed angle (glideslope) to the runway threshold.

The localizer signals are processed by the transceiver and data are transmitted to Air Traffic Control.

The improved version that follows shows you how you can use short sentences and key words to write a text clearly.

Improved version（STE）

Instrument Landing System

The instrument landing system on the aircraft shows data that helps the pilot during the approach to the runway. This system shows the pilot the deviations from the localizer course and the glideslope path. The localizer course is aligned with the centerline of the runway. And the glideslope path is at a constant angle to the threshold of the runway. During the approach to the runway, deviation pointers in the course indicators show the pilot in which direction the aircraft must go:

—Left or right (for the localizer).

—Up or down (for the glideslope).

This data about deviations from the localizer course and glideslope path comes from two VHF transceivers. These transceivers transmit this data to Air Traffic Control.

修改之后的版本有什么特点?

Part V Maintenance English Application Practice

Choose the best answer from the following choices.

( ) 1. There is not too much wear of the rollers or signs of grooves on the flap track.

A. 滚轮上没有过多的磨损且襟翼轨道上没有腐蚀迹象。

B. 滚轮上没有过多的磨损且襟翼轨道上没有划痕迹象。

C. 滚轮上没有过多的磨损且襟翼轨道上没有凹痕迹象。

( ) 2. Apply two layers of protective compound to prevent hydraulic fluid damage to the sealant.

A. 涂上两层保护性化合物，防止液压油损害密封胶。

B. 涂上两层保护密封胶，以防止液压油损害混合物。

C. 在无保护的表面涂上两层密封胶，防止液压油损害化合物。

( ) 3. Put the tools in the solvent to make the compound soft.

A. 在材料表面使用工具（涂抹）使部件变得柔软。

B. 在溶液中使用工具（搅拌）使化合物变得柔软。

C. 在溶剂中使用工具（搅拌）使化合物变得柔软。

( ) 4. The specification contains little new material or novel approaches.

A. 规范中包含一些新的材料和方法。

B. 规范中几乎没有包含什么新材料或新方法。

C. 规范中既不包含新材料，也不包含新方法。

( ) 5. Not all of the material in the specification is applicable to any one model and provision is made for manufacturers to omit material inappropriate to specific aircraft type or model.

A. 本规范中所有的材料都不适用于任何一种型号，并且制造商规定应省略不适用于特定飞机类型或型号的材料。

B．本规范中所有的材料都不适用于任何一种型号，并且制造商应省略不适用于特定飞机类型或型号的材料条款。

C．本规范中并非所有材料都适用于任何一种型号，并规定制造商应省略不适用于特定飞机类型或型号的材料。

（　）6．如果填充物与工具牢固的联合在一起，用冷水湿润工具。

A．If the filter bonds to the tool, make the tool moist with cold water.

B．If the filler bonds to the tool, make the tool moist with cool water.

C．If the filler bonds to the tool, make the tool moist with cold water.

（　）7．化合物腐蚀在表格 5002 中有规定。

A．Corrosion with the chemical compound is specified in table 5002.

B．Corruption with the chemical compound is specified in table 5002.

C．Corruption with the chemical compensation is specified in table 5002.

（　）8．清洁涂抹密封胶的区域。

A．Clean the area that you attach the fluid.

B．Clean the area which you add the agent.

C．Clean the area where you applied the sealant.

（　）9．为了使固化过程更快，给化合物加热。

A．To make the melting process faster, apply heat to the compound.

B．To make the curing process faster, apply heat to the compound.

C．To make the evaporation process faster, apply heat to the compound.

（　）10．这种材料接触热表面时会释放出危险的气体。

A．This material replaces dangerous fumes when it touches hot surface.

B．This material releases dangerous fumes when it touches hot surface.

C．On contact with hot surfaces, this material produces acrid fumes.

（　）11．清洁表面，以确保黏附牢固。

A．Clean the surface to ensure there will be good adhesion.

B．Clean the surface to ensure firm addition.

C．Clean the surface to make sure that the bond is satisfactory.

（　）12．Clean the groove with trichloroethane.

A．用乙烷清洗沟。

B．用三氯乙烷清洗槽。

C．用甲醇清洗槽。

（　）13．Small quantities of surface blooming, which can occur on items in storage, do not cause damage to the component.

A．Small quantities of blooming, do not cause damage to the component, on surface can appear on objects in storage.

B. Those items in storage can be damaged by small quantities of blooming on surface.

C. Small quantities of surface blooming can be found on stored items and do not cause damage to the component.

(　) 14. It is not permitted to use alternative materials.

A. It is fine to use alternative materials.

B. It is forbidden to use alternative materials.

C. The alternative materials are not unserviceable.

(　) 15. Examine the surface.

A. Survey the surface.

B. Do an inspection of the surface.

C. Check the surface.

Part VI Safety and Regulations

Knowledge and Identification of Dangerous Aviation Materials

The supply and guarantee of aviation materials is one of the indispensable links in the operation of airlines, and aviation materials contain more dangerous goods. Therefore, the transportation of hazardous chemical materials is also the focus of the company's dangerous goods transportation safety management.

Names and categories of common dangerous aviation materials.

(1) Non-flammable and non-toxic gas

① Fire extinguishing bottle

② Aviation oxygen cylinder

③ Canned compressed gas

(2) Flammable liquids

① Partial dissolving agent

② Part of the grease

③ Fuel oil

④ Detergent

⑤ Aint

（3）Oxidizer

① AB glue

② Respiratory protection device PBE or passenger service component PSU（including chemical oxygen generator）

③ Oxygen mask

（4）Corrosive substances

① Disinfectant

② Bonding adhesive

（5）Miscellaneous dangerous goods

① Battery（battery）

② Aircraft engines

③ Aircraft auxiliary power equipment

④ Shock absorber pillar

⑤ Fuel pump

⑥ Fuel regulating assembly

⑦ Aircraft emergency slides

⑧ Life jacket

⑨ Tension positioning launcher

Your opinion

Civil aviation safety is of paramount importance. Please talk about the relationship between respecting regulations and civil aviation safety.

Keys for Reference

译文

Module

05

Typical Aircraft Systems

Project 1 Flight Control System

Objectives

To know primary flight control surfaces and secondary flight control surfaces.

To understand the components of flight control system.

To master the usage of cutting tools.

To know Murphy's Law.

Part I Aviation Situational Communication

1. Read aloud the topic-related sentences for aviation communication and try to match them with their Chinese meaning.

________ (1) MCAS refers to Maneuvering Characteristics Augmentation System.

________ (2) Stall is a flight condition during which lift is destroyed due to slow airplane speed or attitude.

________ (3) The angle of attack is excessive.

________ (4) The autopilot is off.

________ (5) The flaps are retracted.

Listen

A. 迎角过大。
B. 襟翼收回。
C. MCAS 是指机动特性增强系统。
D. 自动驾驶仪断开。
E. 失速是因为飞行速度过低或飞机的姿态而引起升力失去的飞行状态。

2. Listen to some short dialogues between A and B. After the dialogues there are some questions. Please choose the most appropriate answer from the choices.

(1) Situation one

Listen

1. What doe MCAS refer to? (   )
A. Marine Corps Air Station (US).
B. Calibrated Air Speed.
C. Maneuvering Characteristics Augmentation System.
D. Mechanical Computer-Aided Design.

2. What is the function of MCAS? (   )
A. Force the nose down.
B. Pitch the nose up.
C. The flaps are retracted.
D. The landing gear is down.

3. What is stall? (   )
A. A landing configuration.
B. A take-off configuration.
C. A flight condition.
D. An approaching configuration.

(2) Situation two

1. When does MCAS activate? (   )
A. When these two conditions are met.
B. When these three conditions are met.
C. When these four conditions are met.
D. When these five conditions are met.

2. Which one is not included in these three conditions? (   )
A. The angle of attack is excessive.
B. The flaps are retracte.
C. The autopilot is off.
D. The autopilot is engaged.

Part II Aviation Reading

2.1 Aviation Reading

Pre-reading Questions

1. What's the purpose of the aircraft flight control systems?

2. What does the primary flight control system consist of?

3. What does the secondary flight control system consist of?

Flight Control System

The aircraft flight control systems allow pilots to control the speed and direction of their aircraft. They govern the necessary inputs to manipulate control surfaces for the pilot to control the aircraft.

Aircraft flight control systems consist of primary and secondary systems. Primary controls are simply those flight controls that the pilot primarily uses to control the airplane. The ailerons, elevator and rudder constitute the primary control system and are required to control an aircraft safely during flight. Secondary flight controls on the other hand are used to change the airplanes performance and lighten the pilots workload. It may consist of wing flaps, leading edge slats, spoilers, horizontal stabilizer and trim systems (Fig. 5-1).

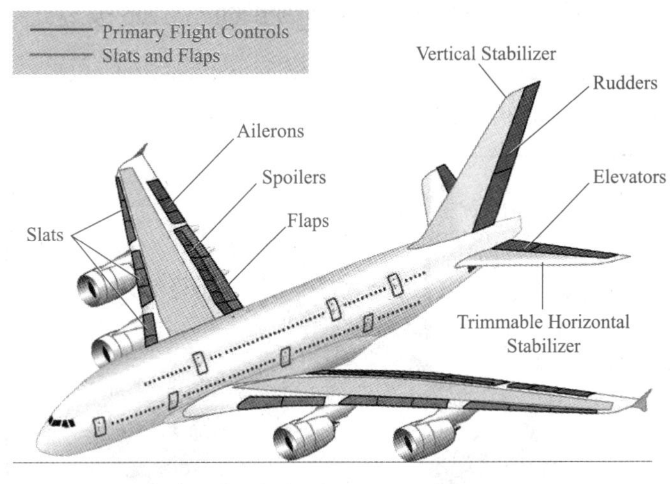

Fig. 5-1 Control surface

Primary Flight Controls

Flight Control System

The three primary flight controls are:

-Ailerons (roll the aircraft left or right).

-Elevator (pitch the aircraft up or down).

-Rudder (yaw the aircraft left or right).

The ailerons are attached to the outboard trailing edge of each wing and move in the opposite direction from each other. They control the airplanes roll about the longitudinal axis.

The elevator is attached to the back of the horizontal stabilizer. It controls the airplanes pitch about the lateral axis, which allows the airplane to climb or descend.

The rudder is a hinged movable surface which is mounted on the after side of the vertical stabilizer. It controls yaw movement of the aircraft about its vertical axis. The rudder is pedal operated by the captain or the first officer (Fig. 5-2).

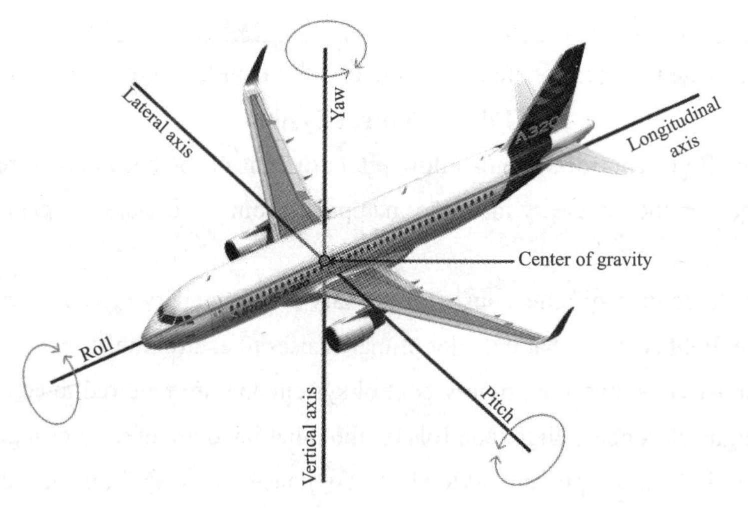

Fig. 5-2 Axis definitions

The pilot controls the ailerons and the elevator with a yoke or stick, while the rudder is controlled through rudder pedals located in the foot well (Fig. 5-3).

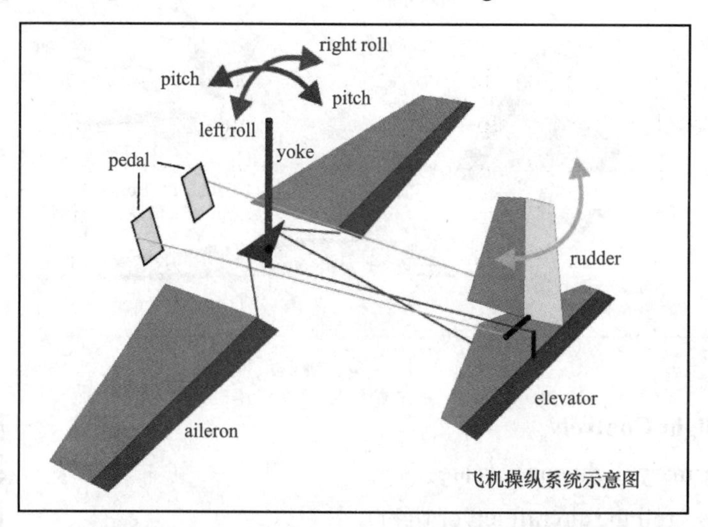

Fig. 5-3 Schematic diagram of flight control system

Secondary Flight Controls

Wing flaps, leading edge slats, spoilers, horizontal stabilizer and trim systems constitute the secondary control system and improve the performance characteristics of the airplane or relieve the pilot of excessive control forces.

The flaps are hinged or pivoted parts of the leading and/or trailing edges for increasing lift at reduced airspeed during landing and take-off. The pilot controls the flaps by moving a lever in the airplane which moves the flaps electrically or manually.

Slats are also high lift devices. They are located on the leading edge of the aircraft wing. Slats extend outwards from the leading edge of the wing. Primarily, all slats increase the camber of the wing. With increased camber comes increased lift. During take-off and landing, the slats can be used to change the angle of attack to produce additional lift（Fig. 5-4）.

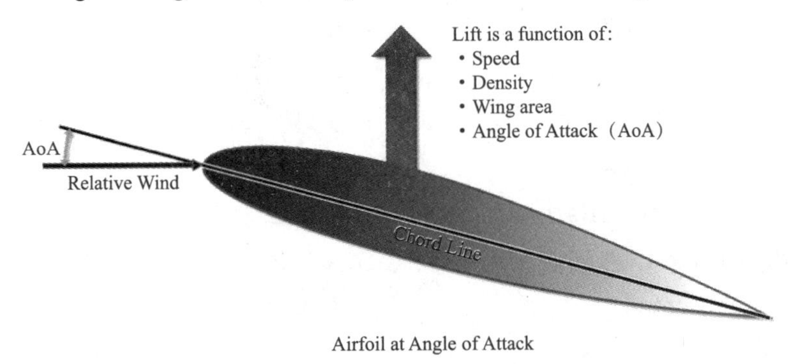

Fig. 5-4 Angle of attack

Found on some fixed-wing aircraft, high drag devices called spoilers are deployed from the wings to spoil the smooth airflow, reducing lift and increasing drag. The spoilers are located on the top of the upper surface of wings. Deploying spoilers on both wings at the same time allows the aircraft to descend without gaining speed. Spoilers are also deployed to help reduce ground roll after landing.

The horizontal stabilizer is a fixed horizontal piece attached to the fuselage of an aircraft. The horizontal stabilizer prevents up-and-down, or pitching, motion of the aircraft nose. As a result, it can provide stability for the aircraft, to keep it flying straight.

Trim is similar to cruise control in the car. It allows pilot to maintain a set altitude by alleviating control pressures. Adjusting trim allows pilot to release that pressure while keeping the airplane at the same altitude. The trim is used to make the pilots job easier and allow the airplane to essentially fly itself with fewer control inputs by the pilot.

In most general aviation airplanes, as the pilot moves the controls, he or she is moving the steel cables or push rods connected through other linkages that physically move these controls. By doing so, the airplane can move about these axes through roll, pitch, and yaw.

Words and Expressions

manipulate	[məˈnɪpjuleɪt]	*v.* 操纵，控制
aileron	[ˈeɪləran]	*n.* 副翼
elevator	[ˈelɪˌveɪtər]	*n.* 升降舵
rudder	[ˈrʌdər]	*n.* 方向舵
workload	[ˈwɜrkˌloʊd]	*n.* 工作量

spoiler	['spɔɪlər]	n. 扰流板
trim	[trɪm]	n. & v. 配平，配平片
flap	[flæp]	n. 襟翼
slat	[slæt]	n. 缝翼
roll	[roʊl]	v. 滚转
pitch	[pɪtʃ]	v. 俯仰
yaw	[jɔː]	v. 偏航
pedal	['ped(ə)l]	n. 脚蹬，踏板
lever	['levər]	n. 控制杆
stability	[stə'bɪləti]	n. 稳定性
primary flight control		主飞行控制
secondary flight control		辅助飞行控制
leading edge		前缘
trailing edge		后缘
fixed-wing aircraft		固定翼飞机
horizontal stabilizer		水平安定面
angle of attack		迎角

Fill in the blanks with the proper words given below, changing the form if necessary.

roll	elevator	pedal	lever	stability

1. Give me a ______ long enough and a fulcrum on which to place it, and I shall move the world.
2. One of our favorite things to do as children was to ______ snowballs down hills.
3. The ______ allows the pilot to pull up.
4. The stabilizers can provide ______ for the aircraft.
5. A rudder ______ are located on the floor of the cockpit.

2.2 Manual Reading

There are some tasks for you to fulfill. You should read the Manual Reading materials carefully and finish the tasks.

Rudder and Rudder Trim Control System—Rudder Pedals and Forward Quadrants

Purpose

The pilots use the rudder pedals to command a yaw control through the rudder forward quadrants.

Location

The rudder pedals are in the flight compartment. The rudder pedal support and forward quadrant assemblies are in the forward equipment compartment.

Physical Description

Each pedal assembly has these components:

* Pedals（2）
* Pedal arm（2）
* Pushrods（2）
* Yoke
* Jackshaft
* Forward quadrant
* Bus rod
* Adjustment crank
* Adjustment flexshaft
* Pedal adjustment nut

Functional Description

The pedals of each pair move in opposite direction to each other. When the pedals move, they move the pushrods and the jackshaft yoke. This moves the jackshaft and the forward quadrant. This also moves the bus rod and makes the other pair of pedals move equally.

The two pairs of rudder pedals can be adjusted independently to suit the captain and first officer. This is accomplished by means of the rudder pedal adjust shaft. The rudder pedal adjustment mechanism consists of an adjustment crank, adjustment shaft, a jackscrew, and pedal adjustment nut attached to the jackshaft assembly.

The adjustment crank is located on the instrument panel forward of the control wheel. The crank is connected to a flex shaft routed forward under the instrument panel, then down under the floor, and aft to the universal joint in the rudder control jackshaft assembly. Rotation of the rudder pedal adjustment crank actuates the jackscrew which causes the yoke, containing the pedal adjustment nut, to move fore and aft.

Rudder pedal adjustment crank and crank handle stops are installed to prevent the rudder pedal adjustment screw from being back driven by heavy foot pressure simultaneously applied to both rudder pedals. The crank incorporates a spring-loaded pin within the knob. The stops are incorporated into the crank housing bearing retainer. Rotation of the crank handle is prevented in either direction beyond the stop blocks, by contact of the spring-loaded pin that protrudes from the crank handle. To permit crank rotation for rudder pedal adjustment, the knob must be pulled aft so the pin clears the stops.

Selected from AMM

Words for Reference

rudder	[ˈrʌdər]	*n.* 方向舵
pedal	[ˈpedl]	*n.* 踏板
quadrant	[ˈkwɑːdrənt]	*n.* 扇形盘
yaw	[jɔː]	*n.* 偏航
assembly	[əˈsembli]	*n.* 配件，装配
pushrod	[pʊʃrəʊːd]	*n.* 推杆
yoke	[joʊk]	*n.* 支架，轭
jackshaft	[ˈdʒækˌʃæft]	*n.* 中间轴，传动轴
crank	[kræŋk]	*n.* 曲柄
flexshaft	[ˈfleksˌʃæft]	*n.* 挠性轴
nut	[nʌt]	*n.* 螺母
adjust	[əˈdʒʌst]	*v.* 调整；调节
jackscrew	[dʒækˌskruː]	*n.* 千斤顶螺旋；制动螺杆
rotation	[roʊˈteɪʃn]	*n.* 旋转；转动
pin	[pɪn]	*n.* 钉，销
knob	[nɑːb]	*n.* 旋钮
protrude	[proʊˈtruːd]	*v.* 突出；伸出

Reading Tasks

1. Judge the following statements according to the contents of the AMM.

(1) The rudder pedals and forward quadrant assemblies are in the flight compartment. (　　)

(2) The first officer could not adjust the rudder pedals under his feet. (　　)

(3) The rudder is controlled through rudder pedals by the pilots. (　　)

2. Fill in the blanks according to the above reading materials.

(1) The two pairs of rudder pedals can be ________ independently to suit the captain and first officer.

(2) ________ of the rudder pedal adjustment crank actuates the jackscrew to move fore and aft.

(3) Rotation of the crank handle is prevented in either direction beyond the stop blocks, by contact of the spring-loaded pin that ________ from the crank handle.

Translate the following sentences into Chinese.

1. The pilots use the rudder pedals to command a yaw control through the rudder forward quadrants.

____

2. The pedals of each pair move in opposite direction to each other.

____

3. The rudder pedal adjustment mechanism consists of an adjustment crank, adjustment shaft, a jackscrew, and pedal adjustment nut attached to the jackshaft assembly.

____

4. The two pairs of rudder pedals can be adjusted independently to suit the captain and first officer.

____

5. Rudder pedal adjustment crank and crank handle stops are installed to prevent the rudder pedal adjustment screw from being backdriven by heavy foot pressure simultaneously applied to both rudder pedals.

____

Part III Tools and Equipments

Cutting Tools

1. Match the words or phrases with the translations.

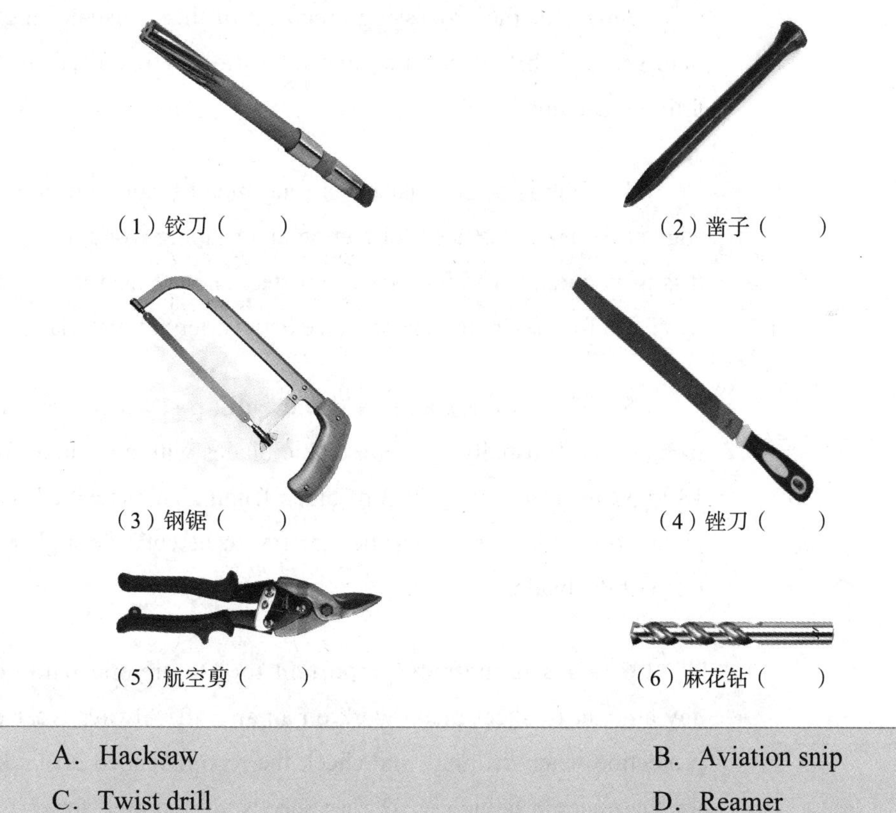

(1) 铰刀 (　　)　　(2) 凿子 (　　)

(3) 钢锯 (　　)　　(4) 锉刀 (　　)

(5) 航空剪 (　　)　　(6) 麻花钻 (　　)

A. Hacksaw	B. Aviation snip
C. Twist drill	D. Reamer
E. Chisel	F. File

2. **Fill in the words or phrases in the blanks.**

Reamer	File	Twist drill
Chisel	Aviation snip	Hacksaw

(1) This is a tool like large scissors, used for cutting metal. The aviation maintenance technician will occasionally fabricate sheet metal parts requiring cutting and trimming.

(2) This is a metal tool with a rough surface for cutting or shaping hard substances or for making them smooth. This cutting tool has rows of teeth shaped like tiny chisels cut diagonally across its face.

(3) This is a kind of finishing tool with a straight edge or spiral edge. It is used to enlarge holes or polish holes. The requirement for the processing precision of this is usually higher than the drill bit. It can be hand operation or installed on the drilling machine.

(4) This is a tool that has a long metal blade with a sharp edge at the end. It is used for cutting and shaping wood and stone. It is usually made of high grade tool steel and are heat treated and tempered for maximum performance on a variety of materials.

(5) This is the standard metal cutting saw used by most technicians. It usually has replaceable blades with anywhere from 18 to 32 teeth for every inch of blade length. This is called blade pitch. Normally, the harder the material to be cut, the higher the pitch of the blade.

(6) This is the most important tool to anyone who does any amount of sheet metal work on an aircraft. Always wear eye protection when drilling, and check the recommended drill speed for the material being cut. Higher speeds are required for drilling soft materials while the lower speeds are used for hard metal such as stainless steel.

Part IV Simplified Technical English

Safety Instructions
安全说明

Definitions 定义

Safety instructions tell the readers that procedures or steps in procedures can be dangerous or cause damage.

安全说明告诉读者程序或程序中的步骤可能是危险的或造成损坏的。

—A warning tells the reader that there is a risk of injury or death.

—A caution tells the reader that there is a risk of damage to objects.

How to write safety instructions 如何编写安全说明

★规则：使用适当的词（如，“warning”或“caution”）来确定风险级别。

Examples in STE:

WARNING: ALWAYS KEEP YOUR HANDS AND FEET AWAY FROM THE BLADE. WHEN THE MOTOR OPERATES, THE BLADE TURNS AND CAN CAUSE INJURY.

CAUTION: BEFORE YOU OPERATE THE GROUND TEST UNIT, MAKE SURE THAT THE PRESSURE REGULATOR IS SET TO ZERO. THIS WILL PREVENT DAMAGE TO THE UNIT.

Examples:

Non-STE: CAUTION:EXTREME CLEANLINESS OF OXYGEN TUBES IS IMPERATIVE.

STE: WARNING: MAKE SURE THAT THE OXYGEN TUBES ARE FULLY CLEAN. OXYGEN AND OIL OR GREASE MAKE AN EXPLOSIVE MIXTURE. AN EXPLOSION CAN CAUSE INJURY OR DEATH TO PERSONNEL.

思考

STE 示例有什么特点?

在非 STE 示例中，本安全说明作为注意事项。但如果了解氧气系统，我们知道氧气会导致爆炸，由于此处存在受伤或死亡的风险，因此必须将此安全说明标识为警告。

Part V Maintenance English Application Practice

Choose the best answer from the following choices.

(　) 1. After the ailerons go back to neutral make sure that they are flush with the flaps.

A. 副翼回到中立位后，确保其与襟翼齐平。

B．扰流板回到原位后，保证和襟翼对齐。

C．在副翼返回中心位之后，保证它们和襟翼相连。

（　）2．Operate the elevator in the full travel 6 times.

A．来回操作升降舵 6 次。

B．操作升降舵 6 个行程。

C．全行程操作升降舵 6 次。

（　）3．确保不要操作相邻的操纵装置。

A．Assure that you do not do the adjacent control.

B．Make sure that you do not operate the adjacent control.

C．Ensure that the adjacent control will be operated.

（　）4．If the rotors stop suddenly, examine the intake for unwanted material.

A．如果襟翼突然停止，检查进气口是否有异物。

B．如果副翼突然停止，检查排气口是否有不必要的材料。

C．如果旋翼突然停止，检查进气口是否有不必要的材料。

（　）5．In these conditions, neither the mechanical nor the electrical pitch trim systems will operate.

A．在这些情况下，俯仰的机械配平和电配平系统都将不工作。

B．当出现以下这些情况，俯仰的机械配平和电配平系统都将继续工作。

C．在这些情况下，电配平系统都将继续工作，俯仰的机械配平不再工作。

（　）6．Put the blanking caps on the open ends of the telescopic duct and the slat duct.

A．在伸缩管和缝翼输送管的开口端盖上堵盖。

B．在伸缩管和缝翼输送管的末端打开堵盖。

C．在防冰管和缝翼输送管的末端打开堵盖。

（　）7．Data shall be presented as indicated airspeed and calibrated airspeed versus flap configuration（any flap position for which performance has been quoted）and angle of bank at maximum weight with throttle.

A．指示空速或校准空速与襟翼配置（任何襟翼配置的性能已被引用），排除最大重量时关闭油门的坡度角表示数据。

B．数据应以指示空速和校准空速，襟翼配置（任何襟翼配置的性能已被引用）减去最大重量时关闭油门的坡度角来表示。

C．数据应以指示空速和校准空速与襟翼配置（任何襟翼配置的性能已被引用）和最大重量时关闭油门的坡度角来表示。

（　）8．飞行员操作手册的本部分应仅包含法规要求的限制或为飞机安全操作所必需并经监管机构批准的限制。

A．This chapter of the pilot's operating handbook shall contain only those limitations required by regulation or necessary for safety operation of the airplane and approved by the regulatory authority.

B. This section of the pilot's operating handbook shall abstain only those limitations required by regulation or necessary for safe operation of the airplane and approved by the regulatory authority.

C. This section of the pilot's operating handbook shall contain only those limitations required by regulation or necessary for safe operation of the airplane and approved by the regulatory authority.

(　) 9. 气流沿着缝翼的顶部和底部表面流动。

A. The airflow then goes along the above and bottom surfaces of the aileron.

B. The air value then goes through the cover and bottom surfaces of the flank.

C. The airflow then goes along the top and bottom surfaces of the slats.

(　) 10. Three computers collect data and parameters from the systems and put them together on the data link to the display unit.

A. Data and parameters, collected by three computers from the systems,are transferred to the display unit through the data link.

B. Data and parameters on the data link are collected and gathered together by three computers from the systems to the display unit.

C. Data and parameters are collected from the systems and gathered together on the data link to the display unit by three computers.

Part VI Safety and Regulations

Murphy's Law

Murphy's Law is commonly expressed as "If anything can go wrong, it will (and at the worst possible time)."

Murphy's Law is usually thought to be named after Captain Edward Murphy, a development engineer with the United States Air Force. In the 1940s and 1950s, he was working with acceleration and deceleration experiments at Edwards Air Force Base. Murphy's Law most likely originated during his projects with Dr. John Paul Stapp.

On February 1, 2003, the US Space Shuttle Columbia broke into pieces over Texas as it returned to the earth, killing all seven astronauts on board. The Columbia space shuttle crash also confirmed Murphy's Law. Such a complex system is bound to fail, either today or tomorrow, which is reasonable.

Murphy's Law is one good reason why engineers always test, test, and test everything. They need to imagine every possible disastrous outcome in order for these outcomes not to be realized. Designers need to take into account all the possibilities that a user will mess up, and somehow make them impossible or at least incapable of causing harm.

Many variants of Murphy's Law exist in today's culture. It is generally accepted, for example, that bread will always land jelly side down when dropped, that there will be rain as soon as you wash your car, and that you will always pick the line in the supermarket that doesn't seem to advance at all.

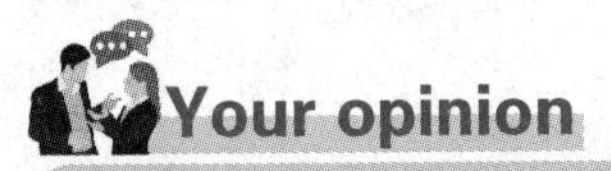

Your opinion

Can you provide some other examples of Murphy's Law in daily life?
What is the inspiration of Murphy's Law for aircraft maintenance personnel?

Project 2 Hydraulic System

Objectives

To know the basic components of hydraulic system.

To understand the three independent hydraulic systems.

To master the usage of pneumatic tools.

To know Hain's Law.

Part I Aviation Situational Communication

1. Read aloud the topic-related sentences for aviation communication and try to match them with their Chinese meaning.

_______ (1) The air pressure is abnormal.

_______ (2) We need to depressurize.

_______ (3) Be careful to protect your hands and your face.

_______ (4) The pressure drops down.

_______ (5) The depressurization valve is closed.

Listen

A. 压力下降了。

B. 小心保护你的手和脸。

C. 我们需要减压。

D. 减压活门关闭。

E. 气压异常。

2. Listen to some short dialogues between A and B. After the dialogues there are some questions. Please choose the most appropriate answer from the choices.

(1) Situation one

1. Which hydraulic system is abnormal? ( )

A. The yellow hydraulic system.

B. The green hydraulic system.

C. The blue hydraulic system.

D. Auxiliary hydraulic system.

2. What should we do to make the air pressure normal? ( )

A. Pressurize this hydraulic reservoir.

B. Compress the air.

C. Depressurize this hydraulic reservoir.

D. Press the airflow.

3. Do they need ground equipment? (   )

A. No, it is not necessary.

B. Yes, a specific tool.

C. Yes, it is necessary.

D. Yes, we can't do it without a tool.

(2) Situation two

1. What do we protect when we make the air pressure normal? (   )

A. Equipment.

B. Depressurization valves.

C. Hydraulic system.

D. Hands and faces.

2. What should we do to make the air pressure normal? (   )

A. Close the manual depressurization valves of the reservoir.

B. Open the manual depressurization valves of the reservoir.

C. Remove the manual depressurization valve.

D. Install the manual depressurization valve.

Listen

Part II Aviation Reading

2.1 Aviation Reading

Pre-reading Questions

1. What is the Chinese meaning of "hydraulic"?

____

2. What is a hydraulic system in an aircraft?

____

3. Why do aircraft use hydraulic systems?

____

Hydraulic System

The word hydraulic is based on the Greek word for water, and originally meant the study of physical behavior of water at rest and in motion. Today the meaning has been expanded to include the physical behavior of all liquids, including hydraulic fluids.

Hydraulics are used in planes of all sizes to operate most of their equipment, such as landing gear, brakes, flaps, thrust reversers and flight controls. Hydraulic systems have many advantages. For instance, they are very reliable and dependable systems, because they require a small number of moving parts, which means lighter weight, ease of installation, simplification of inspection and minimum maintenance requirements. In addition, hydraulic systems are cost-effective. Operations of hydraulic systems are almost 100 percent efficient, with only negligible loss due to fluid friction. They also respond very quickly to control inputs, which is important because it directly affects the efficiency and safety of the airplane.

A typical hydraulic system comprises the hydraulic fluid and a number of components. The basic components are reservoirs, pumps, valves, filters, accumulators, actuators.

Reservoirs

The reservoir works as a storage space for the hydraulic fluid in the hydraulic system. The fluid flows from the reservoir into the system where it performs the required actuations before returning to the reservoir (Fig. 5-5). Many reservoirs have two outlets. One is located in the bottom and the other is connected to a stand pipe that sticks up inside the reservoir (Fig. 5-6).

Fig. 5-5 Hydraulic reservoir filling

Fig. 5-6 Hydraulic oil level gauge

Pumps

A hydraulic system requires a pump to power the system and ensure the hydraulic fluid under pressure is delivered to the actuators when required. Several types of hydraulic pumps driven by a variety of power sources can be found in aviation applications. According to different types of aircrafts, the motive power for the pumps can be generated by a wide variety of options, which generally include manual, engine-driven, electric, pneumatic, hydraulic and Ram Air Turbine (RAT).

Valves

The valves are used in an aircraft hydraulic system to control the flow rate, flow direction, and the pressure in the system. There are numerous types of valves used in aircraft hydraulic systems, generally divided into two categories of flow control valves and pressure control valves. Flow control valves control the rate of flow and the flow direction, but they are not normally concerned with the pressure. Pressure control valves, on the other hand, adjust, regulate, or limit the amount of pressure in the system, or in any portion of the system.

Filters

The filters are the necessary devices in a hydraulic system to ensure the hydraulic fluid is cleaned as it passes around the system. During normal operation, wear and tear of the valves, pumps and other components cause tiny particles of metal to break off and go into suspension in the fluid. These particles must be removed by the filter to prolong the life of the various components and to avoid abrasion.

Accumulators

Aircraft hydraulic accumulators are a vital part of the hydraulic system, which are mainly used to store hydraulic fluid under pressure. Accumulators also provide limited system operation in the event of a failure of the pump (Fig. 5-7).

Fig. 5-7 Accumulator

Actuators

As the fluid pressure is created by the pump and transmitted through the hydraulic system, it arrives at the actuator whose job it is to turn the force of the fluid flow into mechanical power.

Words and Expressions

hydraulic	[haɪˈdrɔːlɪk]	*adj.* 液压的
hydraulic fluid		液压油
installation	[ˌɪnstəˈleɪʃn]	*n.* 安装
reservoir	[ˈrezərvwɑːr]	*n.* (液压)油箱

pump	[pʌmp]	n. 泵 v. 用泵抽送
valve	[vælv]	n. 活门，阀门
filter	['fɪltər]	n. 过滤器 v. 过滤
accumulator	[ə'kjuːmjəleɪtər]	n. 蓄压器，储压器
actuator	['æktʃueɪtər]	n. 作动筒，传动装置
pipe	[paɪp]	n. 管子，管道

Fill in the blanks with the proper words given below, changing the form if necessary.

filter	install	valve	hydraulic	pump

1. The bridge is worked by ______ power, an accumulator with a load of 34 tons supplying pressure water at 630 pounds per square in.
2. ______ of the new system will take several days.
3. The engine is used for ______ water out of the mine.
4. Use a sun block that ______ UVA effectively.
5. The pump sucks air out through the _____ .

2.2 Manual Reading

There are some tasks for you to fulfill. You should read the Manual Reading materials carefully and finish the tasks.

HYDRAULIC POWER—INTRODUCTION

General

There are three independent hydraulic systems that supply hydraulic power for user systems.

The main and auxiliary hydraulic systems supply pressurized fluid to these airplane systems:

* Both thrust reversers

* Power transfer unit (PTU) motor (Fig. 5-8)

Fig. 5-8 Power Transfer Unit

* Landing gear extension and retraction

* Nose wheel steering

* Main gear brake

* Primary flight control

* Secondary flight control

These systems make up the hydraulic power system:

* Main hydraulic systems

* Ground servicing system

* Auxiliary hydraulic systems

* Hydraulic indicating system

Main Hydraulic Systems

The main hydraulic systems are A and B. System A has most of its components on the left side of the airplane and system B on the right side.

Ground Servicing System

The ground servicing system fills all hydraulic reservoirs from one central location.

Auxiliary Hydraulic Systems

The auxiliary hydraulic systems are the standby hydraulic system and the Power Transfer Unit (PTU) system.

The standby hydraulic system is a demand system that supplies reserve hydraulic power to these components:

* Rudder

* Leading edge flaps and slats

* Both thrust reversers

The hydraulic Power Transfer Unit (PTU) system is an alternate source of hydraulic power for the leading edge flaps and slats and auto slat system.

Hydraulic Indicating Systems

These are the indicating systems:

* Hydraulic fluid quantity

* Hydraulic pressure

* Hydraulic pump low pressure warning

* Hydraulic fluid overheat warning

The hydraulic indicating systems show these indications in the flight compartment:

* System A and B reservoir quantity

* Standby reservoir low quantity

* System A and B pressure

* System A and B Engine-Driven Pump (EDP) low pressure (Fig. 5-9)

* System A and B Electric Motor-Driven Pump (EMDP) low pressure

Fig. 5-9　Engine-Driven Pump

* Standby electric motor-driven pump low pressure
* System A and B Electric Motor-Driven Pump（EMDP）overheat

Selected from AMM 29

Words for Reference

auxiliary	[ɔːɡ'zɪliəri]	*adj.* 辅助的；备用的
motor	['moʊtər]	*n.* 发动机，马达
extension	[ɪk'stenʃn]	*n.* 延伸，延长
standby	['stændbaɪ]	*adj.* 备用的
fill	[fɪl]	*v.* 注满，充满
pump	[pʌmp]	*n.* 泵，抽水机
quantity	['kwaːntəti]	*n.* 量，数量
thrust reverser		反推装置
PTU（Power Transfer Unit）		动力转换组件
EDP（Engine-Driven Pump）		发动机驱动泵
EMDP（Electric Motor-Driven Pump）		电动马达驱动泵

Reading Tasks

1. Judge the following statements according to the contents of the AMM.

（1）Only two hydraulic systems can supply hydraulic power for user systems.（　）

（2）The hydraulic power system is composed of main hydraulic systems, ground servicing system, auxiliary hydraulic systems and hydraulic indicating systems.（　）

（3）Most components of hydraulic system A are on the right side of the airplane.（　）

2. Fill in the blanks according to the above reading materials.

（1）The pressurized fluid will be supplied to ______________ from the main and auxiliary hydraulic systems.

（2）All hydraulic reservoirs ________ by the ground servicing system from one central location.

（3）The auxiliary hydraulic systems include the standby hydraulic system and the ____________________ system.

Translate the following sentences into Chinese.

1. There are three independent hydraulic systems that supply hydraulic power for user systems.

____

2. The main and auxiliary hydraulic systems supply pressurized fluid to these airplane systems.

____

3. System A has most of its components on the left side of the airplane and system B on the right side.

____

4. The ground servicing system fills all hydraulic reservoirs from one central location.

____

5. The hydraulic Power Transfer Unit（PTU）system is an alternate source of hydraulic power for the leading edge flaps and slats and auto slat system.

____

Translate the following abbreviations into corresponding Chinese.

1. EDP ____________________

2. PTU ____________________

3. EMDP ____________________

Part III Tools and Equipments

Pneumatic Tools

1. Match the words or phrases with the translations.

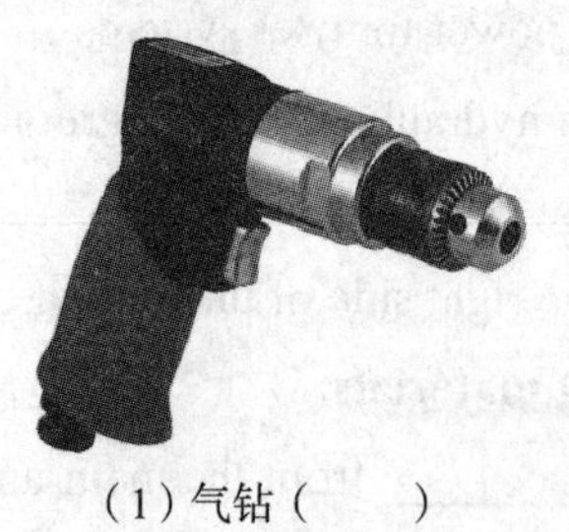

（1）气钻（　　）

（2）气动冲击扳手（　　）

（3）气动铆钉机（　　）

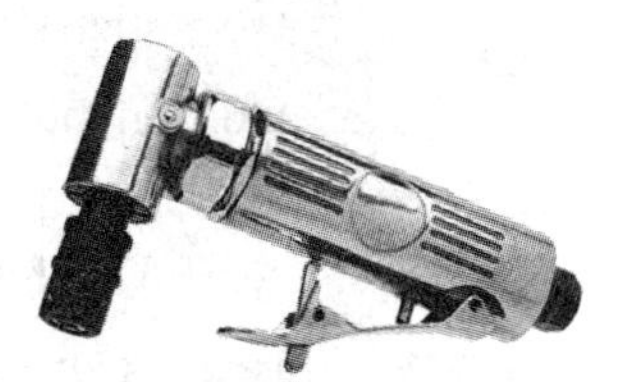

（4）气动打磨机（　　）

（5）气动抛光机（　　）

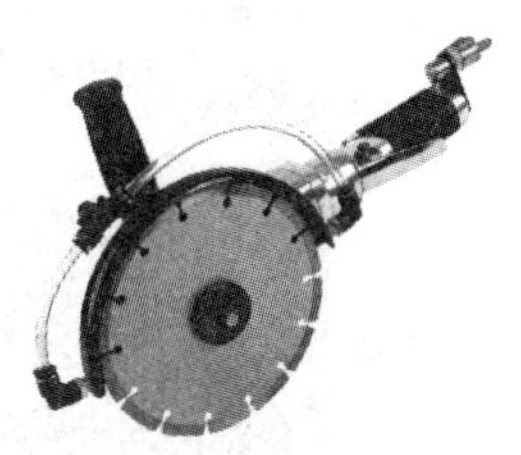

（6）气动切割机（　　）

A. Pneumatic grinding machine	B. Pneumatic cutting machine
C. Pneumatic drill	D. Pneumatic impact wrench
E. Pneumatic polisher	F. Pneumatic riveter

2. Fill in the words or phrases in the blanks.

Pneumatic riveter	Pneumatic grinding machine
Pneumatic impact wrench	Pneumatic drill
Pneumatic polisher	Pneumatic cutting machine

（1）This is a tool that uses air to create enough pressure to drive a rivet into structural metal. There are three basic configurations for this: rivet tools, rivet guns and riveting machines. Each of these works in a similar way, but has a slightly different construction.

（2）This is a pneumatic tool that uses polishing wheels such as cloth and felt to polish the surfaces of various materials. It provides good visual control over the work.

（3）This is suitable for surface grinding of ironwork tire castings, etc. It is suitable for removing the burry edge of finished products in metal processing industry and mould factory, and also suitable for aerospace industry shipyard, etc.

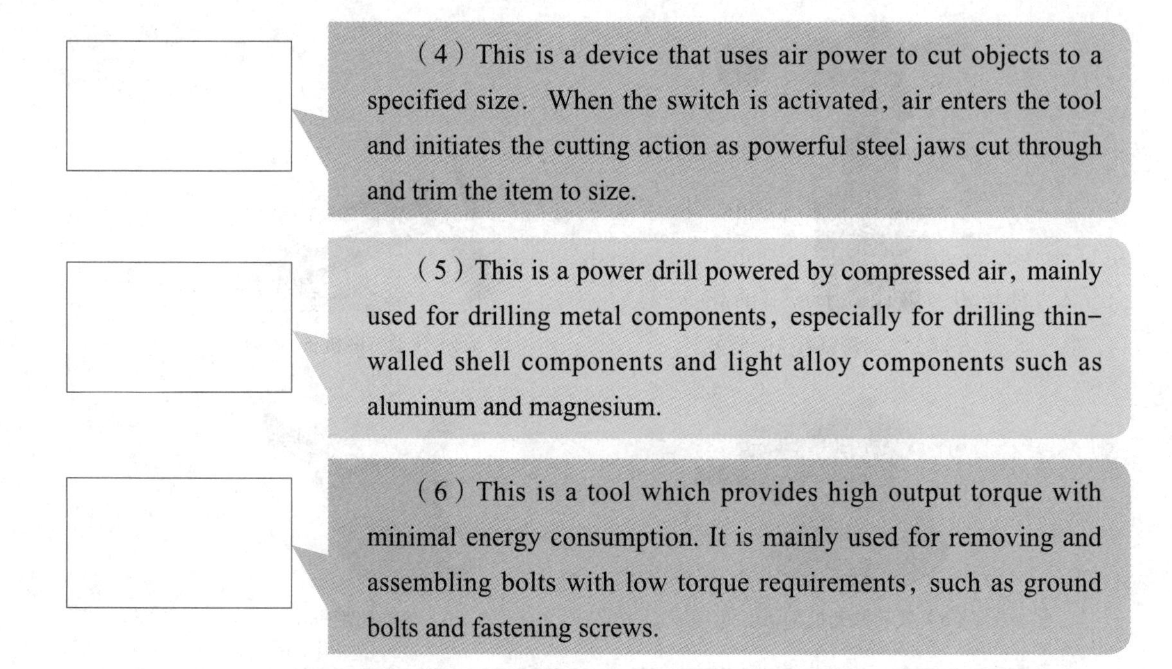

Part IV　Simplified Technical English

Safety Instructions

安全说明

如何编写安全说明

★规则：使用简单、清晰的指令来开始安全说明。

以简单、清晰的命令或条件开始安全说明，让使用者知道如何预防事故和保持高水平的安全。

Examples in STE:

WARNING: DO NOT SWALLOW THE SOLVENT. ALWAYS MAKE SURE THAT YOU KNOW THE SAFETY PRECAUTIONS AND FIRST AID INSTRUCTIONS FOR SOLVENTS. SOLVENTS ARE POISONOUS AND CAN CAUSE INJURY OR DEATH TO PERSONNEL.

CAUTION: DO NOT USE BLEACH OR CLEANSERS THAT CONTAIN CHLORINE TO CLEAN THE UNIT. THESE CLEANING AGENTS CAN CAUSE CORROSION.

如果使用者必须在程序或工作步骤开始之前了解特定条件，请先给出此条件。

Examples in STE:

WARNING: WHILE YOU USE THE SPRAY PAINT, POINT THE SPRAY AWAY FROM YOUR FACE. IT CAN CAUSE INJURY TO YOUR EYES.

CAUTION: WHEN YOU ASSEMBLE THE UNIT, DO NOT LET THE PARTS FALL. IF THEY FALL, PERMANENT DAMAGE CAN OCCUR.

Part V Maintenance English Application Practice

Choose the best answer from the following choices.

( ) 1. If there is no hydraulic power to extend the nose gear, the weight of the gear and the airstream pushes the nose gear into the extended position.

A. 如果没有液压动力放下前起落架，起落架的重量和气流会将其推至放下位。

B. 如果液压动力未将前起落架放下，起落架的重量和重力会将其放下。

C. 液压动力可以放下前起落架，起落架的重量和气流也可以将其推至放下位。

( ) 2. If the pressure supply stops,cancel the test.

A. 如果液压供给停止，则取消测试。

B. 如果压力供给停止，则取消测试。

C. 如果压力供给停止，则测试自动取消。

( ) 3. In this configuration, hydraulic pressure bypasses the valve.

A. 在这种构型中，液压避开此活门。

B. 在这种装置中，液压穿过此活门。

C. 在这种构型中，液压旁通此活门。

( ) 4. Usually, the hydraulic fluid flows into the valve through port a and out through portb.

A. 通常，渗透油从 A 口流入活门，从 B 口流出。

B. 通常，液压油从 A 口流入活门，从 B 口流出。

C. 通常，滑油从 A 口流入活门后流经 B 口。

( ) 5. Apply two layers of protective compound to prevent hydraulic fluid damage to the sealant.

A. 涂上两层保护性化合物，以防止液压油损害密封胶。

B. 涂上两层保护密封胶以防止转压油损害混合物。

C. 在无保护的表面涂上两层密封胶，防止液压油损害化合物。

( ) 6. Make sure that the rate of movement of fuel from the wing tanks to the center tank is equal.

A. 确保燃油从机翼油箱到中央油箱的移动速度相等。

B. 应该使燃油从大翼油箱到中央油箱的传输速率相等。

C. 确保燃油从大翼油箱到中央油箱的传输速率相等。

( ) 7. The hydraulic pressure system supplies pressure for the operation of the flight controls.

A. 液压系统为飞行控制系统的操作提供压力。

B. 飞行控制系统为液压系统的操作提供压力

C. 液压系统与飞行控制系统共同提供压力。

(　) 8. The primary cause of valve failure is contamination of the hydraulic fluid.

A. 液压油污染的主要原因是活门失效。

B. 活门失效的主要原因是液压油污染。

C. 液压油活门失效的主要原因是污染。

(　) 9. 这些操纵面是液压作动的，并且在起飞和降落时展开或收回。

A. These control surfaces are hydraulically supplied and are only extended or retracted during take off and landing.

B. These control surfaces are hydraulically moved and are only extended or retracted during take off and landing.

C. These control surfaces are hydraulically powered and are only extended or retracted during take off and landing.

(　) 10. 接近 1 号液压系统的储压器。

A. Get access to the accumulator for the no.l hydraulic system.

B. Get access to the reservoir for the no.l hydraulic system.

C. Get access to the accumulator for the no.l fluid system.

(　) 11. 泵安装在液压马达后面。

A. The gearbox is installed behind the pneumatic motor.

B. The pump is installed behind the hydraulic motor.

C. The pump is installed behind the pneumatic turbine.

(　) 12. 液压会导致腐蚀。

A. Hydraulic pressure can cause corrosion.

B. Oil fluid caused corrosion.

C. Hydraulic fluid can cause corrosion.

(　) 13. 选择要增压的液压系统。

A. Choose the hydraulic system that you will pressurize.

B. Select the hydraulic system that you will pressurize.

C. Select the pressurized hydraulic system.

(　) 14. 使压力接头适应皮托管端头。

A. Adapt the pressure connection to the pilot head.

B. Adjust the pressure connection to the pilot head.

C. Absorb the pressure connection to the pilot port.

(　) 15. Mechanical power is produced by electric motors that are mainly used to drive pumps, for example, fuel and hydraulic pumps or to drive a valve to the commanded position in the fuel, hydraulic or air conditioning system or to move mechanical parts like doors and cargo containers.

A. You can use electric motors to drive fuel and hydraulic pumps or to drive a valve to the commanded position in the fuel, hydraulic or air conditioning system or to move doors and cargo containers.

B. Electric motors can drive fuel and hydraulic pumps or drive a valve to the commanded position in the fuel, hydraulic or air conditioning system or move mechanical parts like doors and cargo containers to produce mechanical power.

C. Electric motors produce mechanical power and their main functions are to drive fuel and hydraulic pumps, to drive a valve to the commanded position and to move mechanical parts like doors and cargo containers.

Part VI Safety and Regulations

Hain's Law

The Hain's Law is the German inventor of the aircraft turbine, Paps Hain, proposed a law on flight safety in the aviation industry. Hain's Law points out that behind every serious accident, there must be 29 minor accidents, 300 attempted preconditions and 1, 000 potential accidents. Hain's Law can be used to detect and prevent accidents in the safety management of aviation maintenance support.

Aviation maintenance system is a complex cross-linked system of " man-machine-environment" . Maintenance safety management must be built on the basis of science: Scientific maintenance, legal maintenance and careful maintenance. On the basis of scientific organization, the work of maintenance personnel should be carried out in strict accordance with equipment management laws and regulations and various operating procedures, safety regulations and technical standards, and management activities should be strictly regulated and constrained. In the process of maintenance security management, Hain's Law emphasizes:

First, all types of aviation maintenance personnel are required to have the technical and professional level and a high degree of political responsibility to be commensurate with their own work and equipment development, and strictly perform their duties.

Second, aviation maintenance engineering personnel should not only have high maintenance technology, but also pay attention to the ideological and work style construction.

Lastly, aviation maintenance engineering personnel should work quickly and accurately and cooperate closely.

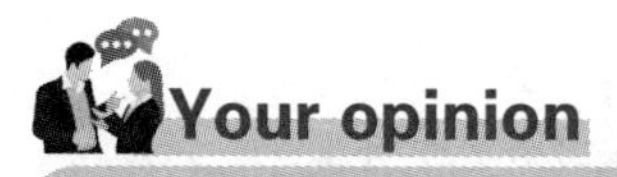

According to Hain's Law, what can aircraft maintenance personnel do to avoid an air crash?

Project 3 Aircraft Fuel System

Objectives

To know the subsystems of aircraft fuel system.

To understand the fuel vent system.

To master the usage of micrometers.

To perceive the Swiss Cheese Model.

Part I Aviation Situational Communication

1. Read aloud the topic-related sentences for aviation communication and try to match them with their Chinese meaning.

Listen

______ (1) There is a fuel leak.

______ (2) Taxi back to the maintenance apron.

______ (3) Our fuel filter light failed.

______ (4) The fuel cap seems to be loose.

______ (5) We'll stop overnight for maintenance.

A. 我们的燃油滤清器灯坏了。

B. 燃料盖似乎松了。

C. 有燃油泄漏。

D. 我们将在这里过夜进行维修。

E. 滑回维修停机坪。

2. Listen to some short dialogues between the Pilot and the Air Traffic Controller. After the dialogues there are some questions. Please choose the most appropriate answer from the choices.

(1) Situation one

1. What problem did CNS123 encounter? ( )

A. Fuel leak on the engine.

B. Fuel leak on the wright wing.

C. Fuel leak on the left wing.

D. Fuel leak on the back wing.

2. Where did CNS123 want to go? ( )

A. The maintenance apron. C. The present position.

B. Fuel leak on the wright wing. D. Fuel leak on the back wing.

3. What did the controller asked CNS123 to do? ( )

Listen

A. Contact tower. C. Backtrack to the maintenance apron.

B. Turn right. D. Taxi back to the maintenance apron.

(2) Situation two

1. What does the crew request? ( )

A. Repair the fuel filter light. C. Replace the fuel filter light.

B. Go back to the ramp. D. Stop overnight for maintenance.

2. Otherwise what would happen to the crew? ( )

A. They'll stop overnight. C. Nothing.

B. They'll stop. D. They'll return back.

Part II Aviation Reading

2.1 Aviation Reading

Pre-reading Questions

1. How many tanks are there in the fuel system of B737NG? And what are they?

____

2. Where can you refuel all tanks?

____

3. Where can you find the fuel quantity on the airplane?

____

Aircraft Fuel System

The airplane is a complex technical object. Like a human or other organisms, it consists of numerous vital systems. The fuel system is one of the most important systems of airplane. It is important part of any vehicle, let alone aircraft, aside from the newest electric powered vehicles.

Aircraft Fuel System

All powered aircraft require fuel on board to operate the engines. The aircraft fuel system stores fuel and delivers the proper amount of clean fuel at the right pressure to meet the demands of the engine or the auxiliary power unit (APU). A well designed fuel system ensures positive and reliable fuel flow throughout all phases of flight, which include changes in attitude, violent maneuvers and sudden acceleration and deceleration. The fuel system consists of these subsystems: Fuel storage system, pressure fueling system, engine fuel feed system, APU fuel feed system, defuel system, fuel quantity indicating system and fuel temperature indication system.

Fuel Storage System

The fuel tanks store fuel for use by the engines and the APU. Using the wet wing concept, all fuel is stored in three tanks, main tank 1, main tank 2, and center tank, which are constructed of the primary wing structure. Front and rear spars, top and bottom wing skins and selected wing ribs are sealed internally to form the tanks. The center tank is in the fuselage and the left and right wing root. The main tanks are located outboard of the center tank. Main tank 1 is in the left wing and main tank 2 is in the right wing. In addition to main and center tanks, surge tanks may also be found on jet transports. These normally empty tanks located in the wing structure outboard of the main wing tanks are used for fuel overflow (Fig. 5-10).

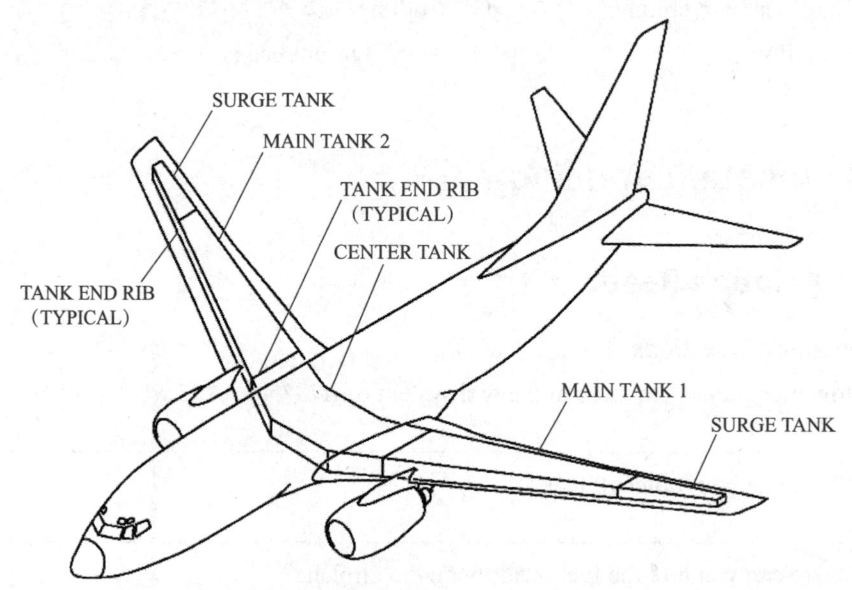

Fig. 5-10 B737NG Fuel Tank Locations

Wing fuel tank access panels permit entry into each fuel and surge tank. Wing ribs divide the fuel tanks into bays. Wing fuel tank access panels are between the wing ribs. Wing rib 8, in main tank 1 and main tank 2, has check valves. The check valves let fuel flow inboard but do not let fuel flow outboard. The sump drain valves let fuel, water, and contamination drain from each fuel tank.

Pressure Fueling System

The pressure fueling system lets you refuel all fuel tanks. Single-point pressure fueling at a fueling station allows all fuel tanks to be filled with one connection of the fuel hose (Fig. 5-11). All tanks fill from the fueling station at the right wing. Various automatic shutoff systems have been designed to close tank fueling valves before the tanks overfill or are damaged. Gauges on the refueling panel allow refueling personnel to monitor progress. In addition, you also use the pressure fueling system during fuel transfer from tank to tank (Fig. 5−12).

Fig. 5−11 Fuel hose

Fig. 5−12 Pressure refueling

Engine Fuel Feed System

The engine fuel feed system is the heart of the fuel system since it delivers fuel to the engines. It supplies fuel to the engines via in−tank fuel boost pumps, usually two per tank. They pump fuel under pressure through a shutoff valve for each engine. A manifold or connecting tubing typically allows any tank to supply any engine through the use of cross−feed valves. The engines use fuel from the center tank before the main tanks. The engines are normally pressure fed from the center tank until the center tank quantity decreases to near zero. They are normally then pressure fed from their respective main tanks. Check valves are located throughout the fuel system to ensure the proper direction of fuel flow and to prevent transfer of fuel between tanks.

APU Fuel Feed System

The APU fuel feed system supplies fuel to the APU. The APU usually receives fuel from main tank 1. However, with use of the fuel boost pump switches, any fuel tank can supply fuel to the APU. If the boost pumps are off, the APU suctions fuel from main tank 1. A DC boost pump supplies fuel to the APU when the center and main tank cannot supply fuel to the left fuel feed manifold.

Defuel System

The defuel system removes fuel from the fuel tanks to the refuel station. It also permits the transfer of fuel between tanks on the ground. The defuel valve is located on the right wing front spar. It connects between the right engine fuel feed manifold and refuel station tubing. There are two ways to defuel the fuel tanks, pressure defuel and suction defuel. You can pressure defuel any tank. You can only suction defuel main tank 1 and main tank 2. You can transfer fuel between any tank. Suction defuel of main tank 1 will occur only if main tank 2 is suction defueled at the same time. When main tank 2 empties, air will be drawn into the manifold and fuel flow will stop (Fig. 5-13).

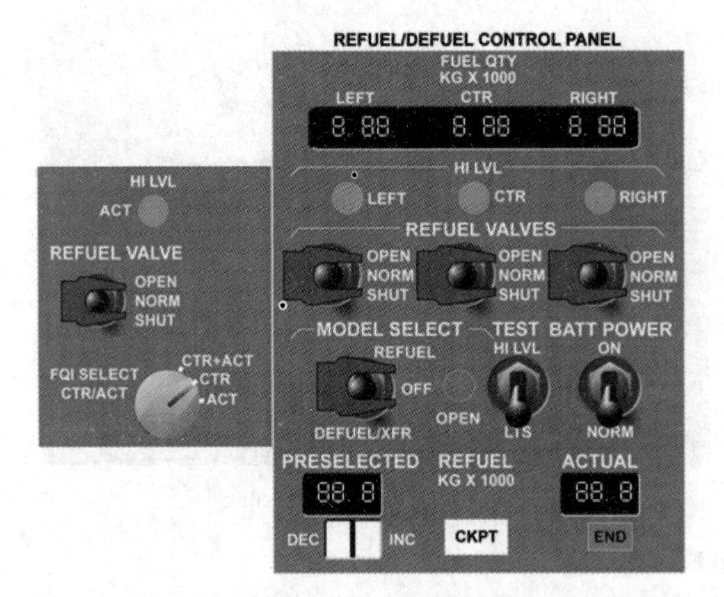

Fig. 5-13 Refuel / defuel control panel

Fuel Quantity Indicating System

The fuel quantity indicating system (FQIS) calculates the fuel weight in each fuel tank. The common display system (CDS) and the fueling panel show fuel quantity. There are 32 tank units in the fuel tanks: There are 12 tank units in main tank 1 and 12 tank units in main tank 2. There are 8 tank units in the center tank. There is one compensator in each tank. Each compensator is in the low point of its fuel tank. The fuel quantity processor unit receives signals from the tank units and the compensators. The fuel quantity processor unit uses these signals to calculate fuel quantity in each tank.

Fuel Temperature Indication System

The fuel temperature indication system shows fuel temperature in main tank 1. The fuel temperature bulb is on the rear spar on main tank 1. It is a resistance unit. The resistance of the fuel temperature bulb changes with fuel temperature. You do not have to defuel main tank 1 to remove the fuel temperature bulb. The fuel temperature indicator is on the fuel system panel (P5-2). It is a resistance ratiometer instrument. The fuel temperature indicator receives 28V AC power and it sends a DC signal through the fuel temperature bulb. The fuel temperature bulb resistance changes this signal before it goes back to the fuel temperature indicator.

Words and Expressions

refuel	[ˌriːˈfjuːəl]	*v.* 加油
defuel	[dɪfˈjuːəl]	*v.* 放油；抽油
subsystem	[səbˈsɪstəm]	*n.* 子系统
sump	[sʌmp]	*n.* 集水坑；集油槽
hose	[həʊz]	*n.* 软管
gauge	[geɪdʒ]	*n.* 计量器
manifold	[ˈmænɪfəʊld]	*n.* 多支管，歧管
respective	[rɪˈspektɪv]	*adj.* 分别的；各自的
compensator	[ˈkɒmpɛnseɪtə]	*n.* 补偿器
bulb	[bʌlb]	*n.* 电灯泡；球茎（状物）
surge tank		通气油箱
check valve		单向活门
cross-feed valve		交输活门
FQIS（Fuel Quantity Indicating System）		油量指示系统
CDS（Common Display System）		公用显示系统
AC（Alternating Current）		交流电
DC（Direct Current）		直流电

Fill in the blanks with the proper words given below, changing the form if necessary.

respective	refuel	defuel	subsystem	gauge

1. The planes needed to ________ before the next mission.
2. Double-check the fuel ________ before taking off.
3. The urban traffic system has three ________.
4. They are each recognized specialists in their ________ fields.
5. The aircraft needs to be ________. Would you please arrange a defuelling-cart?

2.2 Manual Reading

There are some tasks for you to fulfill. You should read the Manual Reading materials carefully and finish the tasks.

Fuel Storage—Fuel Vent System

General

The fuel vent system keeps the pressure of the fuel tanks near the ambient pressure. Too large a pressure difference can cause damage to the wing structure.

Drains let fuel in the vent system return to the tanks.

Flame arrestors make sure excessive heat does not enter the fuel vent system. A clogged flame arrestor causes the pressure relief valve in the surge tank to open. When open, the pressure relief valve becomes another vent for the fuel vent system.

Component Locations

Stringers and the upper wing skin make the vent channels. The vent channels have drain float valves in the center tank.

Vent tubes attach to vent channels. Each vent tube has a drain float valve.

A fuel vent float valve is on the outboard fuel tank end rib in main tank 1 and main tank 2.

A surge tank drain check valve is on the outboard fuel tank end rib in main tank 1 and main tank 2.

The vent scoop (Fig. 5–14) and pressure relief valve are on an access door in each surge tank.

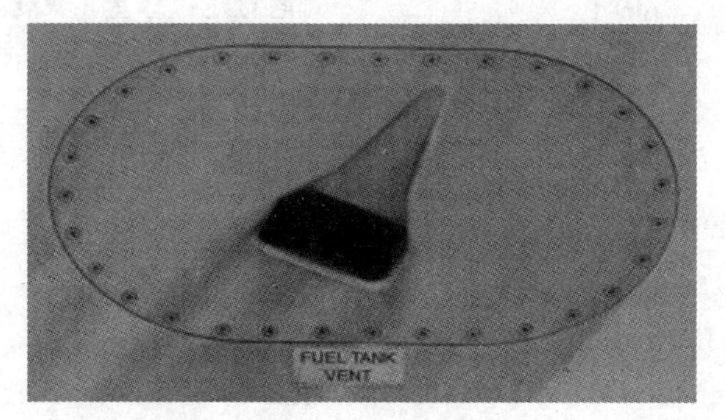

Fig.5–14 Vent Scoop

Functional Description

Vent channels and vent tubes equalize the pressure between each tank and the surge tanks when the airplane is in a climb attitude. The surge tanks are open to the atmosphere through the vent scoop.

The fuel vent float valves equalize the pressure between main tank 1, main tank 2, and the surge tanks when the airplane is in a cruise or descent attitude.

The drain float valves in the vent tubes and the vent channels permit fuel in the vent system to drain into the tank when the fuel level is lower than the valve.

The surge tank drain check valve permits fuel in the surge tank to flow to either main tank 1 or main tank 2. The surge tank drain check valve also prevents fuel flow from main tank 1 and main tank 2 to the surge tank. The pressure relief valve prevents damage to the wing structure when there is too much positive or negative pressure in the fuel tanks. The pressure relief valve is usually closed. When closed, it is even with the bottom surface of the wing. When there is too much positive or negative pressure, the pressure relief valve opens. When it is open, part of the pressure relief valve is in the fuel tank. After it opens, the pressure relief valve stays in the open position. In the open position, the pressure relief valve supplies an additional vent in the surge tank. Pull the reset handle to move the pressure relief valve to the closed position.

For normal operations, make sure the pressure relief valve is closed. An open pressure relief valve is a symptom of a problem in the fuel vent system.

Selected from AMM 28-10

Words for Reference

vent	[vent]	*n.* 通风孔；排气口
stringer	[strɪŋə(r)]	*n.* 桁条
tube	[tjuːb]	*n.* 管子
rib	[rɪb]	*n.* 翼肋
cruise	[kruːz]	*v.* 巡航　*n.* 航行
flame arrestor		阻火器
drain float valve		放泄浮子活门
vent scoop		通气斗
pressure relief valve		释压活门

Reading Tasks

1. Judge the following statements according to the contents of the AMM.

(1) Flame arrestors prevent excessive heat from entering the fuel vent system. (　)

(2) The vent channels are composed of stringers and the upper wing skin. (　)

(3) A fuel vent float valve is on the inboard fuel tank end rib in main tank 1 and main tank 2. (　)

2. Fill in the blanks according to the above reading materials.

(1) The vent channels have drain float valves in the________ tank.

(2) When the airplane is in a cruise or descent attitude, the fuel vent float valves ________ the pressure between main tank 1, main tank 2, and the surge tanks.

(3) In the open position, the pressure relief valve supplies an ________ vent in the surge tank.

Translate the following sentences into Chinese.

1. Too large a pressure difference can cause damage to the wing structure.

____

2. Each vent tube has a drain float valve.

____

3. Vent channels and vent tubes equalize the pressure between each tank and the surge tanks when the airplane is in a climb attitude.

____

4. The drain float valves in the vent tubes and the vent channels permit fuel in the vent system to drain into the tank when the fuel level is lower than the valve.

____

5. The pressure relief valve prevents damage to the wing structure when there is too much positive or negative pressure in the fuel tanks.

____

Part III Tools and Equipments

Micrometers

1. Match the words or phrases with the translations.

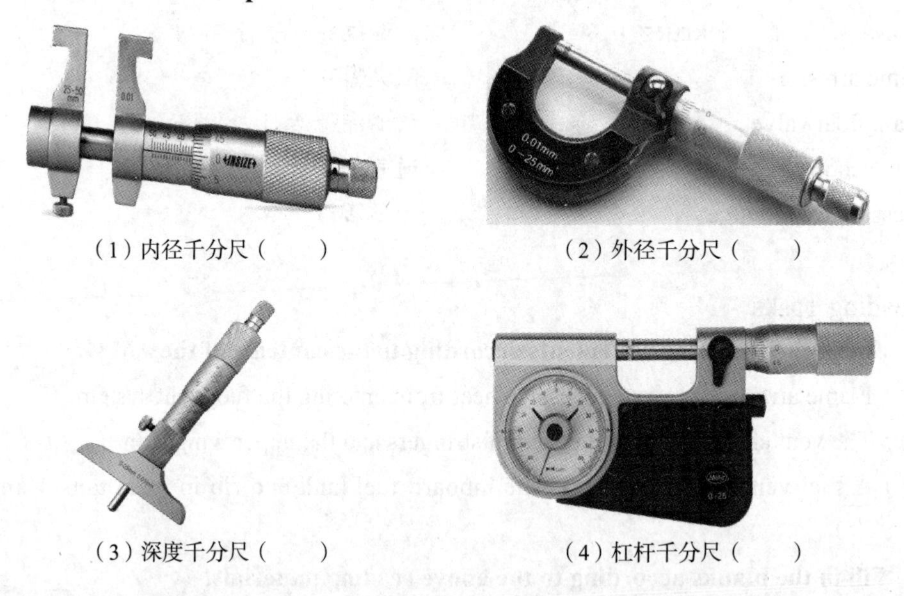

(1) 内径千分尺（　　）　(2) 外径千分尺（　　）

(3) 深度千分尺（　　）　(4) 杠杆千分尺（　　）

A. External micrometer
B. Micrometer with dial comparator
C. Internal micrometer
D. Depth micrometer

2. **Fill in the words or phrases in the blanks.**

Depth micrometer	External micrometer
Micrometer with dial comparator	Internal micrometer

(1) This is an indicating micrometer, using lever gear driving mechanism to read the axial slight displacement. The micrometer consists of micromete head and indicator. It is mainly used for outside measurement and form tolerance measurement.

(2) This is a device designed to precisely measure bore size and common internal dimension. It has a wide measuring range and high precision.

(3) This is a sensitive tool that is used to measure the depth of small holes and bores. It has a ratchet stop. When the measuring head contacts the workpiece to be measured, it keeps stable measuring force.

(4) This is a more precise length measuring instrument than the vernier caliper. The structure of it is composed of a fixed ruler stand, a measuring anvil, a micrometer screw, a fixing sleeve pipe, a differential cylinder, a force measuring device and a locking device.

Part IV Simplified Technical English

Punctuation

标点符号

★规则：可以使用除分号（；)(semicolon）之外的所有标准英语标点符号。

Examples:

Non-STE: (1) Examine the removed parts; replace the damaged ones.

STE: (1) Examine the removed parts for damage.

(2) Replace the damaged part (s).

★规则：使用连字符（-)(hyphens）连接紧密相关的单词。

连字符（-）是连接单词或单词部分的标点符号。

Examples:

1．名词前有两个或两个以上单词和形容词的术语：

low-altitude flight　high-pressure chamber　air-conditioned compartment

transmitter-receiver system　quick-release fastener　fire-resistant material

2．两个单词的分数或数字：

forty-seven　ninety-ninth　one hundred and sixty-two

three-sixteenths　one thirty-second

3．大写字母加名词或数字加名词的术语：

T-shirt　L-shaped bracket　U-beam　Y-coupling

4．以名词或其他词性为第一成分的动词：

stop-drill　vacuum-pack　heat-treat　jump-start　short-circuit

Part V　Maintenance Licence English Application Practice

Choose the best answer from the following choices.

(　) 1．不得触碰液体。

A．Do not permit the fluid to touch you.

B．Do not allow the fluid to touch you.

C．Do not let the fluid to touch you.

(　) 2．A fuel pump is installed in zone 10.

A．三个燃油马达安装在区域 10。

B．一个燃油马达在区域 10 运转。

C．一个燃油泵安装在区域 10。

(　) 3．Hang the shackle on the hoist.

A．把钩环挂在升降机上。

B．用钩环固定升降机。

C．用钩环提起升降机。

(　) 4．This is a recommended precaution to keep the center tank in good condition.It is recommended that you do this procedure at intervals of 100 days.

A．此条为视情采取的预防措施，以使中央油箱保持良好状态。每间隔 100 天视情执行一次此程序。

B．此条为必须采取的预防措施，以使中央油箱保持良好状态。每间隔 100 天视情执行一次此程序。

C．此条为建议采取的预防措施，以使中央油箱保持良好状态。建议间隔 100 天视情执行一次此程序。

(　) 5．The fluid gage includes an 80 cubic centimeters transparent chamber.

A．液位计带有一个 80 立方厘米的透明腔体。

B．液位计带有三个 80 立方厘米的转换腔体。

C．液体比重计带有一个 80 立方厘米的透明腔体。

（　）6．A secondary system is used to estimate the fuel quantity when the aircraft is on the ground.

A．当飞机在地面时用来估计燃油量的辅助系统。

B．当飞机在地面时用来计算燃油量的辅助系统。

C．当飞机在地面时用来探测燃油量的辅助系统。

（　）7．加油软管不能碰到油箱的边缘。

A．The fueling hose must not hit the side of the tank.

B．The fueling hose must not hit the edge of the tank.

C．The fueling hose must not bond the edge of the tank.

（　）8．燃油渗漏会降低发动机性能。

A．Fuel leaks can decrease engine performance.

B．A fuel leak can cause the engine to operate badly.

C．Fuel leakage may be lower engine performance.

（　）9．燃油和氧气的混合物能引起爆炸。

A．A mixture of fuel and oxygen can cause an explosion.

B．A mixture of fuel and hydrogen can cause an explosion.

C．A mixture of fuel and nitrogen can cause an explosion.

（　）10．当飞机加油时，有少量的燃油从通气管流出。

A．Before you refuel the aircraft，a small quantity of oil comes out of the vent line.

B．When you refuel the aircraft，a small quantity of fuel comes out of the vent line.

C．After you refuel the aircraft，a small quantity of fuel comes out of the vent line.

（　）11．计算所需要的燃油量和垫片的数量。

A．Calculate the quality of fuel and the number of shims that are necessary.

B．Calculate the quantity of fuel and the number of shims that are necessary.

C．Calibrate the quantity of fuel and the number of shims that are necessary.

（　）12．让部件的温度下降直到其与周围的温度一致。

A．Let the temperature of the component increase until it is the same as the ambient temperature.

B．Let the temperature of he component decrease until it is the same as the ambient temperature.

C．Let the temperature of the compound decrease until it is the same as the ambient temperature.

（　）13．Move the tube to make the inner connection tight.

A．The intention to move the tube inside is to make tight connection.

B．Remove the tube to achieve a tight inner connection.

C. Move the tube to get the firm inner connection.

(   ) 14. Never exceed speed or mach number is the speed limit that may not be exceeded at any time.

A. Never exceed speed or mach number is the speed limit that can never be exceeded at any time.

B. Never exceed speed or mach number is the speed limitation that will be excessively at any time.

C. Never exceed speed or mach number is the speed limit that may be surpassed at anytime.

(   ) 15. The CAP will address primary and secondary airframe components, and primary and secondary systems to accomplish the stated objective of continued airworthiness.

A. The CAP will cover primary and secondary airframe components, and primary and secondary systems to achieve the stated target of continued airworthiness.

B. The CAP will remark primary and secondary airframe components, and primary and secondary systems to complete the stated objection of continued airworthiness.

C. The CAP will locate primary and secondary airframe components, and primary and secondary systems to accompany the stated objective of continued airworthiness.

Part VI Safety and Regulations

Swiss Cheese Model

The " Swiss Cheese Model ", also known as the " Reason Model ", was put forward by Professor James Reason of University of Manchester in 1990 in his famous psychological monograph *Human Error*.

According to this model, accidents in organizational activities are related to four levels of factors: Environmental impact, unsafe supervision, preconditions of unsafe behavior, and unsafe operational behavior. Each level represents a defense system, and the holes in the level represent loopholes in the defense system. The positions and sizes of these holes are not fixed . The insecurity factor is like an uninterrupted light source. When the holes on each level are in a straight line at the same time, the danger will instantly pass through all the holes like the light source, leading to accidents. These four layers are stacked together like perforated cheese on top of each other, so it is called the "Swiss Cheese Model".

According to the model, a complex system like aviation has multiple layers of security barrier, but there are some " holes " in each layer of security barrier. The " holes " on a single barrier hardly lead to accidents, but when the " holes " on all barriers just form a line, these barriers will be penetrated by unsafe factors, and their safety protection function will fail, thus leading to the occurrence of accidents.

To guard against and defuse security risks is to prevent security risks from being strung together. No matter the size of the " loophole", only by plugging it all can serious consequences be prevented; no matter how serious the security risks are, only when they are completely removed can high risk incidents be prevented.

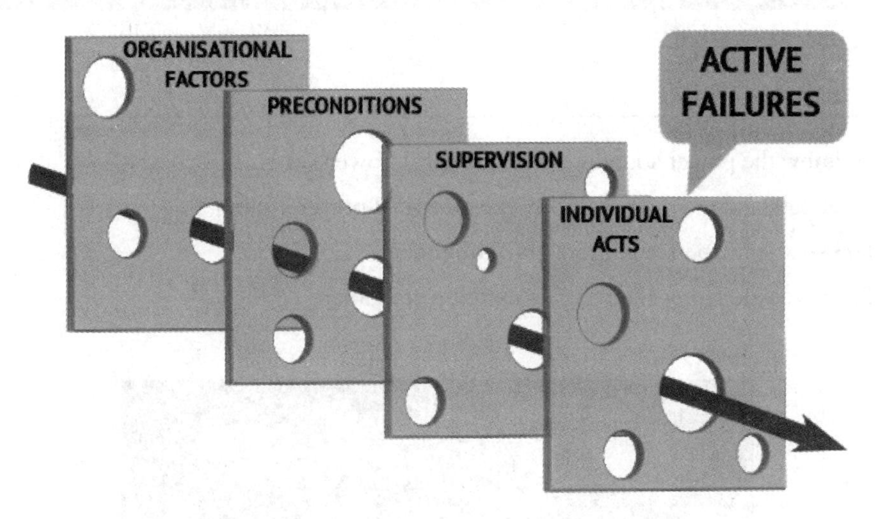

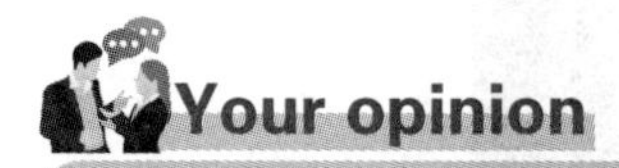

Your opinion

What are the possible reasons for an air crash analyzed by the " Swiss Cheese Model" ?

Project 4　Electrical Power System

Objectives

To know the power sources of the electrical power system.

To understand the distribution of the electrical power system.

To know some personal protective equipment.

To appreciate Hong Jiaguang, a aviation craftsman.

Part I　Aviation Situational Communication

1. Read aloud the topic-related sentences for aviation communication and try to match them with their Chinese meaning.

________ (1) There is dust, lint or hydraulic contamination in the wiring.

________ (2) All the circuits in maintenance are isolated.

________ (3) Supply electrical power to the aircraft.

________ (4) Do a visual inspection of the wiring.

________ (5) Connect electrical equipment to a power source.

Listen

A. 为飞机供电。

B. 导线中有灰尘、纤维屑或液压油污染物。

C. 将电气设备连接到电源。

D. 维修中的所有电路都被隔离了。

E. 对导线进行目视检查。

2. Listen to some short dialogues between A and B. After the dialogues there are some questions. Please choose the most appropriate answer from the choices.

(1) Situation one

1. Which one is not included in a visual inspection of the wiring? (   )

A. ageing   C. overheat

B. deformation   D. over pressure

2. What's meaning of the word "alteration"? (   )

A. 改变   C. 变质

B. 备用   D. 交替

3. What's meaning of the word "contamination"? (   )

A. 沉淀物   C. 有毒物质

B. 污染物   D. 细菌

Listen

(2) Situation two

1. What should be done before supplying electrical power to the aircraft? (   )

A. All the circuits in maintenance are isolated.

B. All the circuits in maintenance are connected.

C. The battery in maintenance is charged.

D. External power is supplied to the airplane.

2. Which sentence is true according to the listening material? (   )

A. Do not connect electrical equipment to a power source less than 30 meters away, unless the power source has not spark-proof connectors.

B. Connect electrical equipment to a power source less than 30 meters away, unless the power source has not spark-proof connectors.

C. Do not connect electrical equipment to a power source less than 30 meters away, unless the power source has spark-proof connectors.

D. Do not connect electrical equipment to a power source more than 30 meters away, unless the power source has not spark-proof connectors.

Part II Aviation Reading

2.1 Aviation Reading

Pre-reading Questions

1. How many main AC power sources does the electrical power system have? And what are they?

2. What can change the 28 V DC into 115 V AC?

---

3. Where can you find the switches to operate the electrical system?

Electrical Power System

New-generation aircraft rely heavily on electrical power because of the wide use of electronic flight instrument systems. The electrical power system makes and supplies AC and DC power to airplane. The system has automatic and manual controls and protection.

There are several different power sources on large aircraft to be able to handle excessive loads for redundancy and for emergency situations.

The electrical power system has four main AC power sources and one standby power source.The main AC power sources include left Integrated Drive Generator (IDG), right IDG, APU starter-generator, and external power (Fig. 5-15). The IDGs and APU starter-generator supply a 3 phase, 115/200 volts at 400 Hz. The IDGs are the primary power source. Each of the engines on an aircraft drives an AC generator. APU starter-generator is the secondary power source. Most often the APU starter-generator is used while the aircraft is on the ground during maintenance or for engine starting. However, most aircraft can use the APU while in flight as a backup power source. External power may only be used with the aircraft on the ground. This system utilizes a Ground Power Unit (GPU) to provide AC power through an external plug on the nose of the aircraft (Fig. 5-16). GPUs may be either portable or stationary units. Major components associated with the AC system include: 3 Generator Control Units (GCUs), Bus Power Control Unit (BPCU) and power panels located in the main equipment center.

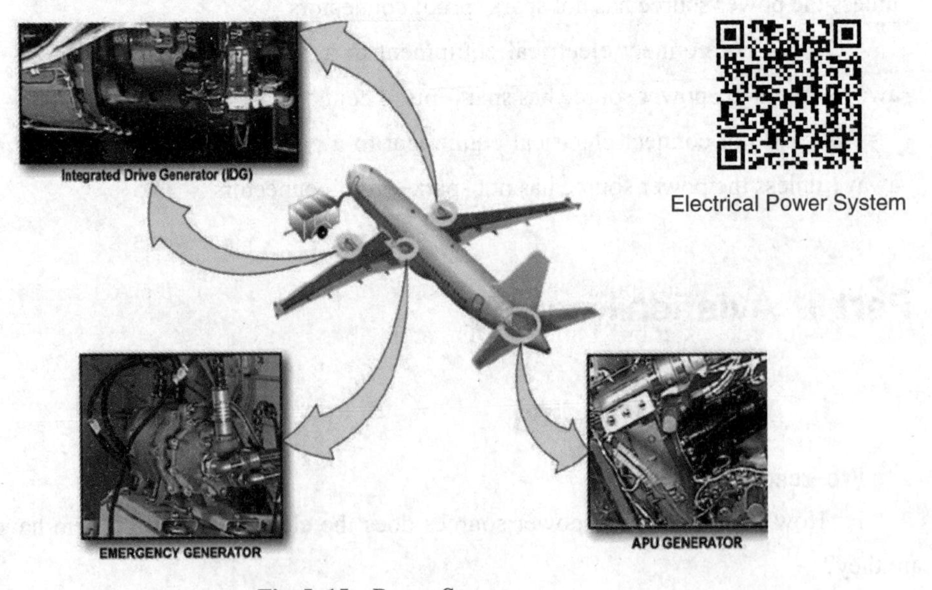

Fig. 5-15 Power Sources

Fig. 5-16 Ground Power

Three Transformer Rectifier Units (TRUs) change 115 V AC to 28 V DC. The DC power sources include main battery, main battery charger, auxiliary battery and auxiliary battery charger. The batteries are the backup DC source if other sources do not operate. The battery provides 28 V DC. It is also possible to change the 28 V DC into 115 V AC with the use of a static inverter. When using the battery, power usage is limited by the short life of the battery.

With the loss of normal power, the standby power system supplies a minimum of 60 minutes of AC and DC power to systems necessary to maintain safe flight. Ram Air Turbine (RAT) may be used, in the case of a generator or APU failure, as an emergency power source (Fig. 5-17) .When necessary, the RAT may be deployed to be used as an AC power source.

There are two basic rules of operation in the electrical power system. The first rule is that AC power sources cannot be operated in parallel. Only one source of AC power can be connected to a transfer bus at a time. And the second rule is that if a second power source is connected to a transfer bus, the first source is automatically disconnected. You use switches on the forward P5 overhead panel or the forward attendant panel to operate the electrical system (Fig. 5-18). The generator drive disconnect switch operates the disconnect mechanism for its IDG. This removes engine accessory gearbox power from the IDG. The engine start lever must be in the idle position for the disconnect function to operate. The standby power switch gives you manual control of the AC and DC standby power bus sources. In the auto position, the AC standby bus receives power from AC transfer bus 1 and the DC standby bus receives power from DC bus 1 when these sources are available. If the sources are not available, the AC standby bus receives power from the static inverter and the DC bus receives power from the battery. You use the ground power switch to control external power to the AC transfer buses. The blue ground power available light above the switch shows if the ground source is connected and quality is good. Both AC transfer buses receive power when you put the ground power switch to the ON position. Any initial power sources are removed before the transfer buses receive external power. You use the ground service switch to supply external power to ground service bus 1 and 2 with external power connected. This makes it possible to supply electrical power for cabin servicing without going into the flight compartment.

Fig. 5-17　Ram Air Turbine

Fig. 5-18　Forward P5 Overhead Panel

Words and Expressions

electrical　[ɪ'lektrɪkl]　*adj.* 电的

excessive　[ɪk'sesɪv]　*adj.* 过分的；过度的

generator　['dʒenəreɪtə(r)]　*n.* 发电机

standby　['stændbaɪ]　*adj.* 备用的　*n.* 备用物品

charger　['tʃɑːdʒə(r)]　*n.* 充电器

phase　[feɪz]　*n.* 阶段；相

stationary　['steɪʃənri]　*adj.* 静止的

portable　['pɔːtəbl]　*adj.* 便携式的

rectifier　['rɛktɪfaɪə]　*n.* 整流器

gearbox　['gɪəbɒks]　*n.* 变速箱

static inverter　静变流机

IDG（Integrated Drive Generator）　整体驱动发电机

GPU（Ground Power Unit）　地面电源装置

GCU（Generator Control Unit）　发电机控制组件

BPCU（Bus Power Control Unit）　汇流条电源控制组件

TRU（Transformer Rectifier Unit）　变压整流器

RAT（Ram Air Turbine）　冲压空气涡轮

Fill in the blanks with the proper words given below，changing the form if necessary.

generator	excessive	portable	standby	electrical

1．There is also a ________ battery in the event the main battery is drained.

2. Solar cells are devices transferring the sun-light to ________ energy.

3. The factory's emergency ________ were used during the power cut.

4. The accident was due to ________ speed.

5. He conceived of the idea to make the first truly ________ computer.

2.2 Manual Reading

There are some tasks for you to fulfill. You should read the Manual Reading materials carefully and finish the tasks.

ELECTRICAL POWER - DISTRIBUTION-GENERAL DESCRIPTION

General

These AC buses receive power directly from an AC power source:

• AC transfer bus 1

• AC transfer bus 2

• Ground service bus 1

• Ground service bus 2

System logic automatically removes loads (load shed) to prevent an overload of an AC power source.

These DC buses receive power directly from the Transformer Rectifier Units (TRUs):

• DC bus 1

• DC bus 2

• Battery bus

These buses receive power directly from the battery or the battery charger:

• Hot battery bus

• Switched hot battery bus

AC Transfer Buses

These AC sources supply power to the AC transfer buses:

• External power

• APU starter-generator

• Integrated Drive Generators (IDGs)

The system design makes sure that two AC power sources can not supply power to the same transfer bus at the same time. However, one AC power source can supply power to both transfer buses through the Bus Tie Breakers (BTBs).

Each transfer bus supplies power to these components or buses:

• Galleys (as many as 2)

• Main bus

• Ground service bus

• Transformer Rectifier Unit（as many as 2）

Ground Service Buses

Each ground service bus receives power in one of these two ways:

• The AC transfer bus on that side has power.

• The ground service switch on the forward attendant panel is in the ON position and external power is connected to the airplane.

The two ground service transfer relays control the selection of the power source.

Main Buses and Galley Buses

The main buses and the galley buses receive power from their respective AC transfer bus. Load shed relays remove the power to these buses when their loads exceed operating limits. This protects the AC power source from overload. The Bus Power Control Unit（BPCU）controls the load shed function.

DC Buses

DC bus 1 usually receives power from TRU 1. However, the bus can receive power from TRU 2 or TRU 3 through the bus tie relay. This relay is usually energized.

DC bus 2 usually receive power from TRU 2. TRU 3 supplies power if TRU 2 fails. DC bus 2 may also receive power from TRU 1 through the bus tie relay.

Standby Buses

The AC standby bus usually receives power from AC transfer bus 1. The static inverter may also supply power to this bus. A Remote Control Circuit Breaker（RCCB）controls power to the static inverter. The DC standby bus usually receives power from DC bus 1. The hot battery bus may also supply power to the DC standby bus.

Battery Buses

The hot battery bus receives power from the battery or battery charger.

The battery bus normally receives power from TRU 3. The battery bus receives power from the battery if TRU 3 has no output.

The switched hot battery bus receives power from the hot battery bus when the battery switch（P5 panel）is in the ON position.

External Power

External power can supply power to these buses:

• AC transfer buses.

• Ground service buses.

External power supplies power to each AC transfer bus through the External Power Contactor（EPC）and the necessary Bus Tie Breaker（BTB）.

APU Power

The APU starter-generator supplies power to each AC transfer bus through the APU breaker（APB）and the necessary BTB. The APU can supply power to both AC transfer buses on the ground or in flight.

IDG Power

The IDGs are the normal power sources of the AC transfer buses. An IDG supplies power through a Generator Control Breaker（GCB）.

Battery Charger

The battery charger makes sure there is maximum battery charge. The battery charger also operates as a TRU when not in the charge mode.

Remote Control Circuit Breaker（RCCB）

The standby power system uses a RCCB to control power input to the static inverter. This RCCB is normally closed.

Forward Attendant Panel

You use the ground service switch to supply external power to ground service bus 1 and 2 with external power connected. This makes it possible to supply electrical power for cabin servicing without going into the flight compartment.

Selected from AMM 24

Words for Reference

bus	[bʌs]	*n.* 汇流条
galley	[ˈgæli]	*n.*（船或飞机上的）厨房
relay	[ˈriːleɪ]	*n.* 继电器
breaker	[ˈbreɪkər]	*n.* 断路器
load shed		减载；切负荷
TRU（Transformer Rectifier Unit）		变压整流器
IDG（Integrated Drive Generator）		整体驱动发电机
BTB（Bus Tie Breaker）		汇流条连接断路器
BPCU（Bus Power Control Unit）		汇流条电源控制组件
RCCB（Remote Control Circuit Breaker）		遥控断路器
EPC（External Power Contactor）		外部电源接触器
APB（APU Breaker）		辅助动力装置断路器
GCB（Generator Control Breaker）		发电机控制断路器

Reading Tasks

1. Judge the following statements according to the contents of the AMM.

（1）The APU can supply power to both AC transfer buses only in flight.（ ）

（2）Both transfer buses can be supplied power by one AC power source through the bus tie breakers（BTBs）.（ ）

（3）It's impossible to supply electrical power for cabin servicing without going into the flight compartment.（ ）

2. Fill in the blanks according to the above reading materials.

(1) Load shed relays remove the power to these buses when their loads ________ operating limits.

(2) The main buses and the galley buses receive power from their ________ AC transfer bus.

(3) The battery charger makes sure there is ________ battery charge.

Translate the following sentences into Chinese.

1. System logic automatically removes loads (load shed) to prevent an overload of an AC power source.

____

2. The system design makes sure that two AC power sources can not supply power to the same transfer bus at the same time.

____

3. The two ground service transfer relays control the selection of the power source.

____

4. DC bus 2 may also receive power from TRU 1 through the bus tie relay.

____

5. The battery charger also operates as a TRU when not in the charge mode.

____

Translate the following abbreviations into corresponding Chinese.

1. IDG ________________

2. TRU ________________

3. GCB ________________

4. RCCB ________________

5. BTB ________________

Part III Tools and Equipments

Personal Protective Equipment

1. Match the words or phrases with the translations.

(1) 工作服（ ）

(2) 安全帽（ ）

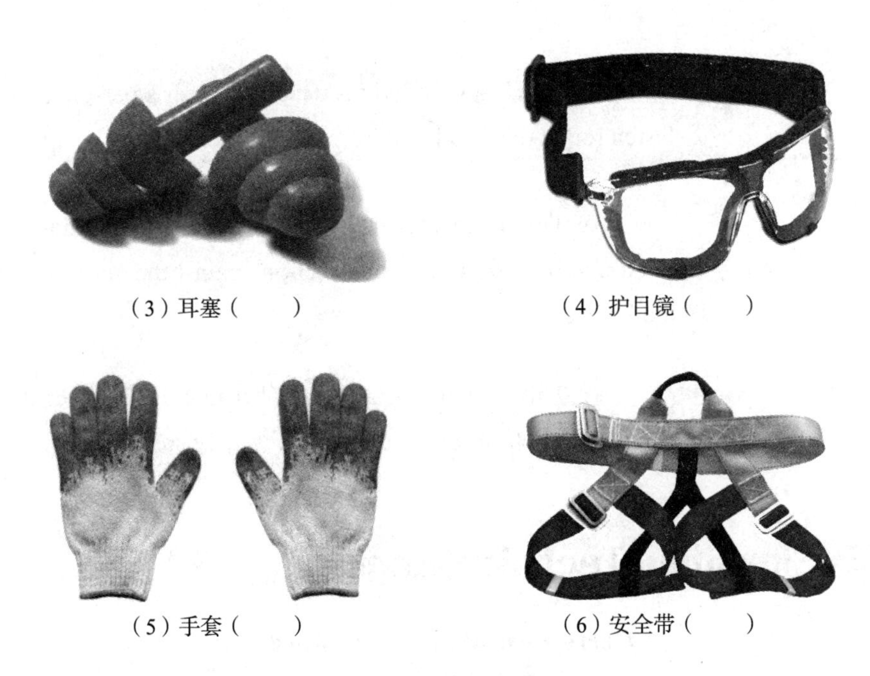

（3）耳塞（　　）

（4）护目镜（　　）

（5）手套（　　）

（6）安全带（　　）

A．Ear plug	B．Gloves
C．Work clothing	D．Safety belt
E．Safety helmet	F．Safety glasses

2. Fill in the words or phrases in the blanks.

Safety glasses	Safety belt
Work clothing	Safety helmet
Gloves	Ear plug

________ (1) They are a type of protective eyewear that usually enclose the eye area, preventing water, chemicals, or particles from getting into the eyes. They are much more resistant than normal prescription glasses.

________ (2) This is a covering designed to be worn on a person's body.

________ (3) They are pieces of clothing which cover your hands and wrists and have individual sections for each finger. You wear them to keep your hands warm or dry or to protect them.

(4) This is a set of straps which fit under a person's arms and fasten round their body in order to to prevent falling or injury.

(5) This is an object made of a soft, pliable material and fitted into the ear canal for protection against the entry of water or loud noise.

(6) This is a strong and hard hat that can protect the head from external force and the injury caused by specific factors.

Part IV Simplified Technical English

Correct use of approved words

正确使用核准词

★规则：正确使用规定的单词。

一些被 STE 认可的词有特定的含义。单词在英语中常常有许多不同的意思。在 STE 中，认可的词通常只有一个特定的含义。该词在标准英语中可能具有的其他含义不予认可。

Examples:

Non-STE: Wear protective clothing.

STE: Use (or put on) protective clothing.

思考

为什么“wear”是非 STE 用法？

“wear”作为动词是不被 STE 认可的，它是一个被 STE 认可的名词，意思是“因摩擦而损坏”。

Non-STE: When the pressure goes down, lift the cover.

STE: When the pressure decreases, lift the cover.

为什么“decrease”是 STE 用法？

“goes down”作为“下降”是一个短语，描述的是当一个指示器，如指针或旗帜下降时的物理状况。“decrease”更好，因为它描述的是压力，而不是监测压力的指示器。

Non-STE: Be careful not to damage the sleeve.

STE: Be careful not to cause damage to the sleeve.

在 STE 使用中，“damage”这个词是被认可的名词，而不是动词。

Please judge which of the following sentences is allowed by STE.

Move the tube to see if the inner connection is tight.

Move the tube to make sure that the inner connection is tight.

Keys:

Non-STE: Move the tube to see if the inner connection is tight.

STE: Move the tube to make sure that the inner connection is tight.

你可以用“see”来形容你能用眼睛看到的东西，但不能用来形容确定或知道。

Part V Maintenance Licence English Application Practice

Choose the best answer from the following choices.

(　) 1. When the meter shows 28 volts DC, the relay closes.

A. 当电表显示 28 伏直流时，断路器闭合。

B. 当电表显示 28 伏直流时，继电器闭合。

C. 当电表显示 28 伏直流时，电磁阀闭合。

(　) 2. For data about the location of circuit breakers, refer to the wiring list.

A. 关于跳开关的位置，参阅线路清单。

B. 关于电路开关的位置，参阅线路清单。

C. 关于电门的位置，参阅线路清单。

(　) 3. Individual wire identification codes are established to assist in tracing circuitry.

A. 建立单独的导线标识码，以组建追踪电路。

B. 建立单独的导线标识码，以协助追踪电路。

C. 建立单独的导线标识码，以装配追踪线圈。

(　) 4. Set the switch to three cycles a minute.

A. 设置电门每秒三个循环。

B. 设置电门每分钟三个循环。

C. 设置电门每三分钟三个循环。

(　) 5. On schematic diagrams such as electrical, where a large number of items are listed, the items shall be presented in a logical order such as the sequence of the arrangement of the items in the airplane or in the schematic diagram.

A. 如果列出了大量项目在电气示意图上，则项目应按逻辑顺序呈现，如飞机上或示意图中项目的排列顺序。

B. 在电气等示意图上，如果列出了大量项目，则项目应按逻辑顺序呈现，如飞机上或示意图中项目的排列顺序。

C. 如果列出了大量项目，例如，电气示意图上的项目应按逻辑顺序呈现，如飞机上或示意图中项目的排列顺序。

( ) 6．这样安装有可能改变了绝缘电缆的走向。

A．It is possible that the installation changed the rotation of the cable loom.

B．It is the installation that changed the routing of the cable loom.

C．It is possible that the sliding of the cable loom is changed by the installation.

( ) 7．系统接口电路适应连接系统的物理特性。

A．The system interconnect circuits adapt to the physical properties of the connected systems.

B．The system interface circuits adapt to the physical protections of the connected systems.

C．The system interface circuits adapt to the physical properties of the connected systems.

( ) 8．设置开关到打开位。

A．Set the switch to on.

B．Move the switch to on.

C．Put the switch to on.

( ) 9．导线必须与接触器后方对接。

A．The wire must touch the front of the contact.

B．The wire must bond the rear of the contact.

C．The wire must touch the rear of the contact.

( ) 10．设置电门到 2 的位置，以得到正确的值。

A．To arrive at the incorrect value，set the switch to position 2.

B．To get the correct value，set the switch to position 2.

C．To get the incorrect value，set the switch to position.

( ) 11．Individual electrical symbols are do not have an established code.

A．This specification，for those who do not have an established code，individual electrical symbols are given as a suggestion.

B．As a suggestion for those who do not have an established code，individual electrical symbols are gained in this specification.

C．As a suggestion for those who do not have an established code，individual electrical symbols are required in this specification.

( ) 12．Adjust the potentiometer until you do not hear a hum from the loudspeaker.

A．If there is hum from the loudspeaker，don't adjust the potentiometer.

B．Adjust the potentiometer unless hear a hum from the loudspeaker.

C．Adjust the potentiometer until the loudspeaker no longer hums.

( ) 13．Do not let the cable touch the floor and make sure the floor is not wet.

A．Do not Keep the cable hit the floor and make sure the floor is dry.

B. Let the cable above the floor and assure the floor is moist.

C. Do not let the cable touch the floor and make sure the floor is moist.

( ) 14. Individual electrical symbols are provided in this specification as a suggestion for those who do not have an established code.

A. As a suggestion for those who do not have an established code, individual electrical symbols are gained in this specification.

B. In this specification, for those who do not have an established code, individual electrical symbols are given as a suggestion.

C. As a suggestion for those who do not have an established code, individual electrical symbols are required in this specification.

( ) 15. Each electrical system has a code to identify it.

A. Each electrical system is identified by a code.

B. Each code of electrical system is identified.

C. Each system is identified by an electrical code.

Part VI Safety and Regulations

Aviation Craftsman

Hong Jiaguang

" Gold is not inherently there. It is out from thousands of gravel and it is difficult to find gold in the garbage inside. " Remembering his mother's words since he was young, Hong Jiaguang, an ordinary Chinese young man who was born in Shenyang and did farm work since childhood, has now grown into the first " sander " in China, winner of the second prize of National Science and Technology Progress Award in 2017, a representative of the 20th National Congress of the Communist Party of China, and a senior technician of AECC. Hong Jiaguang has participated in many national key aero-engine scientific research projects such as the carrier aircraft of the Liaoning, and has 7 national invention and utility model patents. The most important part is that he conquered the difficult processing problem of diamond roller forming surface: His skill is excellent. The tool he developed for processing aeroengine blades can process blades accurately to 1/80th of a hair, which successfully broke the blockade of Western countries on Chinese aeroengine technology and won honor for China.

Aviation craftsmen are the elite of the nation, the model of the people, and the meritorious officer of the aviation industry of the People's Republic of China.

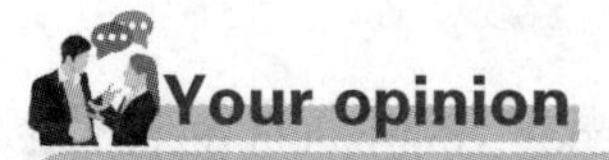

Do you know any other aviation craftsmen? And what have you learned from them?

Keys for Reference

译文

Appendix Ⅰ Selected Vocabulary References for Aviation Maintenance English Testing

Word (part of speech) 单词(词性)	Approved meaning/ Alternatives 已批准含义/可替换单词	Approved example 批准示例	Not approved 未批准
		A	
admit (*v.*) 允许 (*v.*)	LET (*v.*) 使得 (*v.*)	OPEN THE VALVE TO LET NITROGEN GO INTO THE OLEO STRUT. 打开活门，让氮气进入减震支柱。	Open valve to admit nitrogen to the oleo strut. 打开活门，允许氮气进入减震支柱。
adopt (*v.*) 采取、接受 (*v.*)	USE (*v.*) 利用 (*v.*)	USE THIS PROCEDURE IF THE UNIT IS DAMAGED. 如果该组件损坏，请使用此程序。	Adopt this procedure if the unit is damaged. 如果该组件损坏，请使用此程序。
advance (*n.*) 前进、向前 (*n.*)	FORWARD (*adj.*) 向前的 (*adj.*)	THE FORWARD MOVEMENT OF THE CONTROL LEVER MUST BE SLOW AND CONTINUOUS. 控制杆的前移必须缓慢而连续。	The advance of the control lever must be gradual. 控制杆的前移必须是渐进的。
advance (*v.*) 前进、向前 (*v.*)	SET (*v.*) 设置 (*v.*)	SET THE THROTTLE TO MAXIMUM POWER. 设置油门杆到最大功率。	Advance the throttle to maximum power. 向前推油门杆到最大功率。
—	FORWARD (*adv.*) 向前 (*adv.*)	MOVE THE LEVER FORWARD. 向前推杆。	Advance the lever. 向前推杆。
adverse (*adj.*) 不利的 (*adj.*)	BAD (*adj.*) 不好的 (*adj.*) NOTE: Give accurate and correct conditions if possible. 注：如果可能给出精确的或正确的条件	REFER TO CHAPTER 6 FOR INSTRUCTIONS ON HOW TO PARK IN BAD WEATHER CONDITIONS. 关于如何在恶劣天气下停放，请参阅第六章。	For parking aircraft in adverse weather conditions, refer to Chapter 6. 关于在恶劣天气下飞机的停放，请参阅第六章。
advisable (*adj.*) 适当的 (*adj.*)	RECOMMEND (*v.*) 建议 (*v.*)	WE RECOMMEND THAT YOU TORQUE THE BOLTS AGAIN AFTER 50 FLIGHT HOURS. 我们建议你在飞行 50 小时后再次拧紧螺栓。	It is advisable to retorque the bolts after 50 flight hours. 最好在飞行 50 小时后再次拧紧螺栓。
advise (*v.*) 建议 (*v.*)	TELL (*v.*) 通知 (*v.*)	TELL PERSON B THAT THE BRAKES ARE SET. 通知人员设置刹车。	Advise Person B that the brakes have been set. 通知人员设置刹车。

续表

Word (part of speech) 单词（词性）	Approved meaning/ Alternatives 已批准含义 / 可替换单词	Approved example 批准示例	Not approved 未批准
—	RECOMMEND (*v.*) 建议 (*v.*)	WE RECOMMEND THAT YOU REFER TO THE REPAIR MANUAL. 我们建议你遵循修理手册执行工作。	We advise you to consult the repair manual. 我们建议你遵循修理手册。
affect (*v.*) 影响 (*v.*)	EFFECT (*n.*) 影响 (*v.*) NOTE: Be specific if possible. 注：如果可能的话要具体一点	MAGNETIC TOOLS HAYE AN UNWANTED EFFECT ON THE COMPASS SYSTEM. 磁性工具对罗盘系统有不利影响。	Magnetic tools affect the compass system. 磁性工具影响罗盘系统。
AFT (*adj.*) 后部的 (*adj.*)	Nearer to the rear of an air or sea vehicle 靠近空中或海上运输工具的后部	THE PUMP IS IN THE AFT CELL OF THE FUSELAGE TANK. 泵安装在油箱的后部。	—
AFT (*adv.*) 后部地 (*adv.*)	In the direction of the rear of an air or sea vehicle 在空中或海上运输工具的后部	MOVE THE THROTTLE AFT. 将油门杆向后移动。	—
AFTER (*conj.*) 在之后 (*conj.*)	That follows a specified time, sequence, or operation 跟随制定的时间、顺序或操作	DO A FUNCTIONAL TEST AFTER INSTALL THE COMPONENT. 安装部件后进行功能测试。	—
AFTER (*prep.*) 在之后 (*prep.*)	That follows a specified time, sequence, or operation 跟随制定的时间、顺序或操作	THE BAR MOVES DOWN AFTER 20 SECONDS. 20 秒后将杆向下移。	—
AFT OF (*prep.*) 后部 (*prep.*)	At a position nearer to the rear 在靠近后方的位置	THE CONTROL UNIT IS INSTALLED AFT OF THE FLIGHT COMPARTMENT. 控制组件安装在驾驶舱的后部。	—
AGAIN (*adv.*) 再一次 (*adv.*)	One more occurrence 再一次发生	MOVE THE CONTROL STICK BACK TO THE CENTER, THEN MOVE IT FORWARD AGAIN. 移动控制杆回到中间位，然后再向前移动。	—
AGAINST (*prep.*) 紧靠 (*prep.*)	In contact with 与……接触	PUT THE ADAPTER IN POSITION AGAINST ITS SUPPORT. 将适配器紧靠在支架上。	—
AGENT (*n.*) 试剂 (*n.*)	One of a group of materials made to do a specified task 用来完成一项特定任务材料	DO NOT USE THESE CLEANING AGENTS ON HOT SURFACES. 不要在热表面使用这些清洁试剂。	—

Word (part of speech) 单词（词性）	Approved meaning/ Alternatives 已批准含义 / 可替换单词	Approved example 批准示例	Not approved 未批准
aggravate (*v.*) 加剧 (*v.*)	INCREASE (*v.*) 增加 (*v.*)	TIRE WEAR INCREASES WITH HIGH SPEED. 轮胎磨损随速度的提高而增加。	Tire wear is aggravated by high speed. 轮胎因高速行驶而磨损加剧。
AGREE (*v.*), （AGREES, AGREED） 一致 (*v.*)	To be consistent with 与……一致	THE INDICATIONS MUST AGREE WITH THE VALUES IN THE TABLE. 指示必须与表格中的值一致。	—
		B	
begin (*v.*) 开始 (*v.*)	START (*v.*) 开始 (*v.*)	INCREASE THE PRESSURE UNTIL THE FLOW STARTS AGAIN. 增加压力，直到再次开始流动。	Increase pressure until the flow begins again. 增加压力，直到再次开始流动。
beginning (*n.*) 开始 (*n.*)	START (*n.*) 开始 (*n.*)	YOU CAN GET SLOW MOVEMENT AT THE START. 开始要慢慢地移动。	You can get slow movement at the beginning. 开始要慢慢地移动。
—	SOURCE (*n.*) 源 (*n.*)	FIND THE SOURCE OF THE FUEL LEAKAGE. 找到燃油渗漏源。	Find the beginning of the fuel leak. 找到燃油渗漏源。
—	START (*v.*) 开始 (*v.*)	WHEN THE LEVER STARTS TO MOVE, THE MICROSWITCH OPERATES. 当手柄开始移动时，微动电门启动。	At the beginning of the lever movement, the microswitch will operate. 在手柄开始移动时，微动电门将启动。
BELOW (*prep.*) 在……下面 (*prep.*)	In (or to) a position farther down than something 处于（或）比某物更低的位置 NOTE: For other meanings, use: 用于	THE DATE IS WRITTEN BELOW THE CYLINDER NECK. 数据写在缸体颈部以下。	—
—	LESS THAN 少于	MAKE SURE THAT THE DIAMETER OF THE HOLE IS LESS THAN THE SPECIFIED VALUE. 确保孔的直径小于规定的值。	Make sure that the diameter of the hole is below the specified value. 确保孔的直径低于规定的值。
BEND (*n.*) 弯曲 (*n.*)	The area where something is bent 弯曲的区域	EXAMINE THE BENDS FOR CRACKS. 检查弯曲处是否有裂纹。	—
BEND (*v.*), (BENDS,BENT) 弯曲 (*v.*)	To change or cause to change from straight to curved 改变或引起，由直变弯	BEND THE PIPE CAREFULLY. 小心地弯曲管子。	—

Word (part of speech) 单词（词性）	Approved meaning/ Alternatives 已批准含义 / 可替换单词	Approved example 批准示例	Not approved 未批准
—	—	THESE PARTS CAN EASILY BEND, BREAK OR BECOME INCORRECTLY ALIGNED. 这些部件很容易弯曲、断裂或对不齐。	—
beneath (*prep.*) 在……之下 (*prep.*)	BELOW (*prep.*) 在……之下 (*prep.*)	PUT THE JACK BELOW THE AXLE. 把千斤顶放在轮轴下面。	Put the jack beneath the axle. 把千斤顶放在轮轴下面。
beside (*prep.*) 在旁边 (*prep.*)	ADJACENT TO 邻近的	THE FUEL PUMP IS ADJACENT TO THE SPAR. 燃油泵靠近大翼梁。	The fuel pump is beside the spar. 燃油泵靠近大翼梁。
beware (*v.*) 当心 (*v.*)	BE CAREFUL 小心	BE CAREFUL OF DANGEROUS VOLTAGES. 小心危险的电压。	Beware of dangerous voltages. 当心危险的电压。
beyond (*prep.*) 超过 (*prep.*)	MORE THAN 超过	REPLACE COMPONENTS THAT ARE WORN MORE THAN THE MAXIMUM LIMITS. 更换磨损超标的组件。	Replace components that are worn beyond their maximum limits. 更换磨损超标的组件。
		C	
chance (by chance)(*n.*) 可能性 (*n.*)	RISK (*n.*) 风险 (*n.*)	IF THERE IS A RISK OF LEAKAGE, PUT A CONTAINER BELOW THE UNIT. 如果有渗漏的风险，在组件下面放一个容器。	If there is a chance of leakage, put a container below the unit. 如果有渗漏的可能，在组件下面放一个容器。
—	ACCIDENTALLY (*adv.*) 意外地、偶然地 (*adv.*)	IF THE LEVER MOVED ACCIDENTALLY, DO THE TEST AGAIN. 如果手柄意外移动，再做一次测试。	If by chance the control lever has been moved, do the test again. 如果手柄可能被移动了，再做一次测试。
CHANGE (*n.*) 改变 (*n.*)	That which occurs when something changes 当某物发生改变时发生的变化	THE COLOR CHANGE SHOWS THAT THE TEMPERATURE IS TOO HIGH. 颜色的改变说明温度太高了。	—
CHANGE (*v.*), (CHANGES, CHANGED) 改变 (*v.*)	To become or to cause to become different 变得不同或使变得不同	IF THE HUMIDITY CHANGES FREQUENTLY, PUT A COVER ON THE UNIT. 如果湿度变化频繁，请在组件上盖上盖子。	—
—	NOTE: For other meanings, use: 用于	CHANGE THE COLOR OF THE DISPLAY. 改变显示器的颜色。	—

Word (part of speech) 单词（词性）	Approved meaning/ Alternatives 已批准含义 / 可替换单词	Approved example 批准示例	Not approved 未批准
—	REPLACE (*v.*) 替代 (*v.*)	REPLACE THE DAMAGED VALVE. 替换损坏的活门。	Change the damaged valve. 换掉损坏的活门。
channel (*v.*) 输送、引导 (*v.*)	CHANNEL (*n.*) 通道 (*n.*)	INSTALL THE WIRES IN THE CHANNEL ALONG THE HOUSING. 沿着外壳在通道中安装导线。	Channel the wires along the housing. 沿着外壳铺设导线。
characteristic (*n.*) 特性 (*n.*)	PROPERTY (*n.*) 性能 (*n.*)	THE PROPERTIES OF THESE SEALANTS PREVENT CORROSION. 这些密封剂的性能可以防止腐蚀。	The characteristics of these sealants prevent corrosion. 这些密封剂的特性可以防止腐蚀。
—	QUALITY (*n.*) 品质、质量 (*n.*)	DO NOT USE THIS MATERIAL BECAUSE IT DOES NOT HAVE THE NECESSARY QUALITIES. 不要使用这种材料，因为它不具备必需的品质。	Do not use this material because it doesn't have the right characteristics. 不要使用这种材料，因为它不具备正确的特性。
CHARGE (*v.*), (CHARGES, CHARGED) 充电 (*v.*)	To accumulate or add electrical energy 积聚或增加电能	CHARGE THE BATTERY. 给电瓶充电。	—
—	NOTE: For other meanings, use: 用于	MAKE SURE THAT THE BATTERY CHARGES. 确保电瓶充电。	—
—	FILL (*v.*) 使充满 (*v.*)	FILL THE TANK WITH 10 LITERS OF METHANOL. 给油箱装满 10 升甲醇。	Charge the tank with 10 liters of methanol. 向油箱加 10 升甲醇。
—	PRESSURIZE (*v.*) 增压 (*v.*)	PRESSURIZE THE ACCUMULATOR WITH NITROGEN. 用氮气给储压器增压。	Charge the accumulator with nitrogen. 给储压器充氮气。
CHECK (*n.*) 检查 (*n.*)	The procedure you do to make sure that something operates correctly 保证某些操作正确的程序	DO A CHECK OF THE HYDRAULIC SYSTEM. 做液压系统的检查。	—
check (*v.*) 检查 (*v.*)	MAKE SURE (*v.*) 确保 (*v.*)	MAKE SURE THAT IT IS SAFE TO SUPPLY ELECTRICAL POWER. 确保供电是安全的。	Check that it is safe to apply electrical power. 检查用电安全。

Word (part of speech) 单词（词性）	Approved meaning/ Alternatives 已批准含义 / 可替换单词	Approved example 批准示例	Not approved 未批准
—	MEASURE (*v.*) 测量 (*v.*)	MEASURE THE BETWEEN THE FACES. 测量（两个）表面之间的距离。	Check the distance between the faces. 检查（两个）表面之间的距离。
—	EXAMINE (*v.*) 检查 (*v.*)	EXAMINE THE CASTING FOR CORROSION. 检查铸件是否有腐蚀。	Check the casting for corrosion. 检查铸件是否有腐蚀。
—	CHECK (*n.*) 检查 (*n.*)	DO A LEAKAGE CHECK OF THE VALVE. 做活门的渗漏检查。	Check the valve for leakage. 检查活门的渗漏性。
CHEMICAL (*adj.*) 化学的 (*adj.*)	Related to a chemical 与化学相关的	REMOVE THE CORROSION WITH THE CHEMICAL COMPOUND THAT IS SPECIFIED IN TABLE 6001. 用表 6001 中规定的化学化合物去除腐蚀。	—
CHEMICALLY (*adv.*) 化学地 (*adv.*)	Related to a chemical 与化学相关	REMOVE CORROSION CHEMICALLY. 用化学方式去除腐蚀。	—
chip (*n.*) 碎片、碎屑 (*n.*)	PARTICLE (*n.*) 颗粒 (*n.*)	EXAMINE THE FILTER ELEMENT FOR METAL PARTICLES. 检查过滤元件是否有金属颗粒。	Examine the filter element for metal chips. 检查过滤元件是否有金属碎屑。
chip (*v.*) 打破 (*v.*)	DAMAGED (*adj.*) 损坏的 (*adj.*)	IF THE ENAMEL IS DAMAGED, REPLACE THE UNIT. 若釉质受损，则更换该组件。	If the enamel is chipped, replace the unit. 若釉质受损，则更换该组件。
chock (*v.*) 用楔子垫阻 (*v.*)	CHOCK (TN) 轮挡 (TN)	PUT THE CHOCKS AGAINST THE WHEELS. 用轮挡靠紧机轮。	Chock the wheels. 用楔子垫挡住机轮。
		E	
ELECTRICALLY (*adv.*) 有关电地 (*adv.*)	Related to or operated by electricity 相关电或由电操作	THE SYSTEM IS HYDRAULICALLY OPERATED AND ELECTRICALLY CONTROLLED. 系统是电控液动。	—
efflux (*n.*) 流出 (*n.*)	EXHAUST (*n.*) 排气管 (*n.*)	MAKE SURE THAT THERE ARE SAFETY BARRIERS AROUND THE ENGINE EXHAUST AREA. 确保发动机排气管区域围绕有安全栅栏。	Ensure that there are safety barriers around the engine efflux area. 确保发动机排气管区域围绕有安全栅栏。

Word (part of speech) 单词(词性)	Approved meaning/ Alternatives 已批准含义/可替换单词	Approved example 批准示例	Not approved 未批准
effort (*n.*) 努力 (*n.*)	FORCE (*n.*) 力量 (*n.*)	DO NOT TIGHTEN THE BOLTS WITH TOO MUCH FORCE. 不要用太大力拧紧螺栓。	Do not tighten the bolts with too much effort. 拧紧螺栓不要太用力。
either (*adj.*) 两者择一的 (*adj.*)	ONE OF THE TWO 两个中的一个	IF THERE IS MERCURY CONTAMINATION IN ONE OF THE TWO COMPARTMENTS, CLEAN THE AREA IMMEDIATELY. 如果两个舱中有一个舱被水银污染，立即清洁该区域。	If there is mercury spillage in either compartment, clean the area immediately. 如果两个舱中有一个舱中有水银溢出，立即清洁该区域。
either (*pron.*) 或者，要么 (*pron.*)	ONE OF THE TWO 两个中的一个	APPLY ELECTRICAL POWER TO ONE OF THE TWO SOLENOIDS. 向两个电磁阀的其中一个供电。	Apply electrical power to either of the solenoids. 供电到两个中的其中一个电磁阀。
EJECT (*v.*), (EJECTS, EJECTED) 喷射 (*v.*)	To move or to cause a person or item to move from an aircraft or equipment with force 用力使人或物从飞机或设备上移动	IF YOU PULL THE EJECTION SEAT HANDLE, THE SEAT WILL EJECT. 如果拉动弹射座椅手柄，座椅将被弹射。	—
EDGE (*n.*) 边缘 (*n.*)	A line that is the intersection of two surfaces of a solid object 实体的两个表面相交的线	THE DISTANCE BETWEEN THE EDGE OF THE PANEL AND THE PARTITION MUST NOT BE MORE THAN 0.05 mm. 面板边缘与隔板之间的距离不得大于 0.05 毫米。	—
EFFECT (*n.*) 效果 (*n.*)	The result of a cause 致使的结果	WHEN DUST MIXES WITH OIL, IT HAS AN ABRASIVE EFFECT. 当粉末和油混合，就有了研磨的效果。	—
effect (*v.*) 产生，达到目的 (*v.*)	DO or other command verb construction 做或其他命令动词结构	DO THE TIGHTENING PROCEDURE. 执行拧紧步骤。	Effect the tightening procedure. 执行拧紧步骤。
effective (*adj.*) 有效的 (*adj.*)	GOOD (*adj.*) 好的 (*adj.*)	THIS MATERIAL GIVES GOOD PROTECTION FROM CORROSION. 此材料起到了良好的防腐作用。	This material provides effective protection from corrosion. 此材料提供了有效的防腐。

Word (part of speech) 单词（词性）	Approved meaning/ Alternatives 已批准含义 / 可替换单词	Approved example 批准示例	Not approved 未批准
efficacious (*adj.*) 有效的 (*adj.*)	GOOD (*adj.*) 好的 (*adj.*)	THIS IS A GOOD PROCEDURE TO REMOVE PAINT. 这是除漆的好方法。	This is an efficacious way to remove paint. 这是除漆的有效方法。
efficient (*adj.*) 有效率的 (*adj.*)	SATISFACTORY (*adj.*) 满意的 (*adj.*)	THE TRANSMISSION CONTROL PROTOCOL IS NOT SATISFACTORY FOR THE TRANSMISSION OF INTERACTIVE TRAFFIC. 传输控制协议不能满足交互通信的传输。	The Transmission Control Protocol is not efficient for the transmission of interactive traffic. 传输控制协议不能有效用于交互通信的传输。
efficiently (*adv.*) 有效地 (*adv.*)	SATISFACTORILY (*adv.*) 令人满意地 (*adv.*)	MAKE SURE THAT THE UNIT OPERATES SATISFACTORILY. 确保组件工作正常。	Make sure the unit operates efficiently. 确保组件有效运行。
ease (*n.*) 容易 (*n.*)	EASILY (*adv.*) 容易地 (*adv.*)	MAKE SURE THAT YOU CAN MOVE THE HANDLE EASILY. 确保你能轻松移动手柄。	The handle must be moved with ease. 手柄能被轻松移动。
	EASY (*adj.*) 容易的 (*adj.*)	MAKE SURE THAT IT IS EASY TO MOVE THE HANDLE. 确保能轻松移动手柄。	The handle must be moved with ease. 手柄能被轻松移动。
ease (*v.*) 小心缓缓地移动 (*v.*)	CAREFULLY REMOVE 小心移除	CAREFULLY REMOVE THE TRIM COVER FROM THE ADHESIVE TAPE. 从胶带上小心移除装饰罩。	Ease trim cover from adhesive tape. 从胶带上小心移除装饰罩。
	CAREFULLY MOVE 小心移动	CAREFULLY MOVE THE PIPE INTO THE CORRECT POSITION. 小心移动管路进入正确位置。	Ease the pipe into the right position. 小心缓缓地移动管路进入到正确位置。
	G		
get off (*v.*) 离开 (*v.*)	MOVE OFF (*v.*) 离开 (*v.*)	BEFORE YOU MOVE THE AILERONS, TELL ALL PERSONS TO MOVE OFF THE WINGS. 在移动副翼之前，需要告诉所有人离开机翼区域。	Before you move the ailerons, tell all persons to get off the wings. 在移动副翼之前，需要告诉所有人离开这个机翼区域。
get to (*v.*) 到达，开始 (*v.*)	BE (*v.*) 有，发生 (*v.*)	THE MOTOR STOPS WHEN THE FLAPS ARE AT THE END OF THEIR TRAVEL. 当襟翼到达行程终点时，马达停止。	The motor stops when the flaps get to the end of their travel. 当襟翼到达行程终点时，马达停止。

Word (part of speech) 单词（词性）	Approved meaning/ Alternatives 已批准含义 / 可替换单词	Approved example 批准示例	Not approved 未批准
—	GO (*v.*) 到达，去 (*v.*)	GO TO THE AFT CARGO COMPARTMENT TO DO THIS TEST. 到后货舱做这个测试。	Get to the aft cargo compartment to do this test. 到后货舱做这个测试。
—	INCREASE (*v.*) 增加 (*v.*)	WHEN THE TEMPERATURE OF THE VENTILATION AIR INCREASES TO 27 ℃ , THE DISCHARGE VALVE OPENS. 当通风空气温度达到 27 ℃时，排放活门打开。	When the temperature of the ventilation air gets to 27 ℃ , the discharge valve opens. 当通风空气温度达到 27 ℃时，排放活门打开。
GIVE (*v.*), GIVES, GAVE, GIVEN 提供，给 (*v.*)	To provide 提供	THIS SECTION GIVES YOU THE PROCEDURES FOR THE DISASSEMBLY. 本节提供分解的程序。	—
give rise to (*v.*) 引起，造成 (*v.*)	CAUSE (*v.*) 引起 (*v.*) 原因 (*n.*)	SEAWATER CAN CAUSE CORROSION. 海水能引起腐蚀。	Seawater can give rise to corrosion. 海水能引起腐蚀。
gleam (*v.*) 闪烁，发亮 (*v.*)	SHINY (*adj.*) 光滑的 (*adj.*)	POLISH THE SURFACE WITH A SOFT CLOTH UNTIL IT BECOMES SHINY. 用一块柔软的布擦亮表面直到它光滑。	Polish the surface with a soft cloth until it gleams. 用一块柔软的布擦亮表面直到它发亮。
glitch (*n.*) 故障 (*n.*)	ERROR (*n.*) 错误 (*n.*)	IF THERE IS AN ERROR IN THE SYSTEM, THE SCREEN SHOWS: “NO GO” . 如果系统出现错误，屏幕显示：“禁止进入”。	If there is a glitch in the system, the screen shows: NO GO. 如果系统出现故障，屏幕显示：禁止进入。
—	FAILURE (*n.*) 故障，失效 (*n.*)	IF THERE IS FAILURE IN THE SYSTEM, THE SCREEN SHOWS: “NO GO” . 如果系统出现故障，屏幕显示：“禁止进入”。	If there is a glitch in the system, the screen shows: NO GO. 如果系统出现故障，屏幕显示：禁止进入。
—	UNSERVICEABLE (*adj.*) 不工作的 (*adj.*)	IF THE SYSTEM IS UNSERVICEABLE, THE SCREEN SHOWS: “ NO GO” . 如果系统不工作，屏幕显示：“禁止进入”。	If there is a glitch in the system, the screen shows: NO GO. 如果系统出现故障，屏幕显示：禁止进入。
gloss (*n.*) 光泽 (*n.*)	SHINY（*adj.*） 光滑的，光亮的（*adj.*）	POLISH THE SURF ACE UNTIL IT IS VERY SHINY. 抛光表面直到非常光亮。	Polish the surface to a high gloss. 抛光表面直到有高光泽度。

Word (part of speech) 单词（词性）	Approved meaning/ Alternatives 已批准含义 / 可替换单词	Approved example 批准示例	Not approved 未批准
GLOSSY（adj）（GLOSSIER, GLOSSIEST）光泽的（adj）（光泽度更高，光泽度最高）	Smooth and shiny 光滑和光泽	APPLY THE PATCH WITH THE MATT SIDE AGAINST THE GLOSSY SURFACE. 将亚光（粗糙）的一面贴在光泽度高的表面上。	—
glow（V）发光（V）	BE (*v.*) 变 (*v.*)	TURN THE POTENTIOMETER UNTIL THE LIGHT IS DIM. 转动电位器，直到灯变暗。	Turn the potentiometer until the light glows dimly. 转动电位器，直到灯变暗。
glue (*v.*) 粘接 (*v.*)	BOND (*v.*) 结合 (*v.*)	BOND THE PATCH TO THE SURFACE WITH THE APPLICABLE GLUE. 用适当的胶把补片粘在表面上。	Glue the patch to the surface. 把补片粘接到表面上。
—	ATTACH (*v.*) 贴上 (*v.*)	ATTACH THE PROTECTIVE PLATE TO THE SURFACE WITH ADHESIVE MATERIAL. 用粘接材料将保护板贴到表面。	Glue the protective plate to the surface. 把保护板粘接到表面上。
GO (*v.*),（GOES, WENT, GONE）去 (*v.*)	To move to or from something 移动到某物或从某物移动	MAKE SURE THAT THE POINTER GOES OUT OF VIEW. 确认指针消失。	—
GO OFF (*v.*),（GOES OFF, WENT OFF, GONE OFF）停止运转 (*v.*)	To become dark when an internal power source is deenergized 当内部电源断电时变暗	THE ANNUNCIATOR LIGHT GOES OFF. 通告灯熄灭。	—
GOOD (*adj.*) (BETTER, BEST) 好的 (*adj.*)（更好，最好）	That is satisfactory 令人满意的	THIS MATERIAL GIVES GOOD PROTECTION FROM CORROSION. 这个材料提供了很好的防腐保护。	—
gouge (*v.*) 沟槽 (*v.*) 用半圆凿子挖 (*v.*)	GOUGE (*n.*) 沟槽，切口 (*n.*)	IF A GOUGE OCCURS IN THE BLADE DURING REMOVAL, REPLACE THE BLADE. 如果叶片在拆卸过程中有沟槽，请更换叶片。	If the blade was gouged during removal, replace it. 如果叶片在拆卸过程中有沟槽，请更换叶片。
		H	
heighten (*v.*) 提高，增加 (*v.*)	INCREASE (*v.*) 提高，增加 (*v.*)	STRONG WINDS WILL INCREASE THE RISK OF DAMAGE. 强风会增加损坏的风险。	Strong winds will heighten the risk of damage. 强风会增加损坏的风险。

续表

Word (part of speech) 单词（词性）	Approved meaning/ Alternatives 已批准含义 / 可替换单词	Approved example 批准示例	Not approved 未批准
help (*n.*) 帮助 (*n.*)	AID (*n.*) 帮助 (*n.*)	GET MEDICAL AID IMMEDIATELY. 立即到医疗救助。	You must obtain medical help as soon as you can. 你只要尽快就可得到医疗救助。
HELP (*v.*), (HELPS, HELPED) 帮助 (*v.*)	To make something easier or better 使一些事情变得容易或者更好	PETROLATUM HELPS PREVENT CORROSION OF THE TERMINALS. 凡士林有助于防止终端的腐蚀。	—
helpful (*adj.*) 有用的，有帮助的 (*adj.*)	HELP (*v.*) 帮助 (*v.*)	RECORD THE LOCKWIRE POSITIONS, THIS WILL HELP YOU DURING THE ASSEMBLY PROCEDURE. 记录熔丝的位置对于组装的过程是有用的。	Record the lockwire positions. This will be helpful during the assembly procedure. 记录熔丝的位置对于组装的过程是有用的。
HERE (*adv.*) 在这里 (*adv.*)	In this position 在这个位置	TO DOWNLOAD THE FILE, CLICK HERE. 点击这里下载文件。	—
hesitation (*n.*) 暂停，犹豫 (*n.*)	SMOOTHLY (*adv.*) 光滑地，平稳地 (*adv.*)	MAKE SURE THAT THE SOLENOID OPERATES SMOOTHLY. 确认电磁阀工作平稳。	Make sure that the solenoid operates without hesitation. 确认电磁阀工作平稳。
—	CORRECTLY (*adv.*) 正确地，恰当地 (*adv.*)	THE VALVE MUST OPERATE CORRECTLY. 活门必须正确地操作。	The valve must operate without hesitation. 活门必须操作正常。
—	IMMEDIATELY (*adv.*) 立即地，直接地 (*adv.*)	WHEN YOU OPEN THE CIRCUIT BREAKER, THE LIGHT MUST COME ON IMMEDIATELY. 当拔出跳开关灯必须立即点亮。	When you open the circuit breaker, the light must come on without hesitation. 当拔出跳开关，灯必须立即点亮。
HIGH (*adj.*) (HIGHER, HIGHEST) 高级的 (*adj.*)(较高的，最高的)	That is of large value 有很大价值	USE THE SPECIAL PROTECTION FOR STORAGE IN HIGH TEMPERATURES. 在高温下储存，需要使用特殊的保护。	—
highly (*adv.*) 高度地 (*adv.*)	VERY (*adv.*) 很，非常 (*adv.*)	TOLUENE IS VERY FLAMMABLE. 甲苯是非常易燃的。	Toluene is highly flammable. 甲苯是高度易燃物。

Word (part of speech) 单词(词性)	Approved meaning/ Alternatives 已批准含义/可替换单词	Approved example 批准示例	Not approved 未批准
hinder (*v.*) 阻碍 (*v.*)	PREVENT (*v.*) 预防，阻止 (*v.*)	SCRATCHES CAN PREVENT THE FREE MOVEMENT OF THE PISTON IN THE SLEEVE. 划痕将会阻止活塞在套筒中自由移动。	Scratches can hinder the movement of the piston in the sleeve. 划痕将会阻碍活塞在套筒中自由移动。
—	BLOCKAGE (*n.*) 堵塞 (*n.*)	MAKE SURE THAT THERE IS NO BLOCKAGE IN THE PIPE THAT PREVENTS AIRFLOW. 确认管道中没有堵塞阻碍气流。	Make sure that there is nothing in the pipe to hinder airflow. 确认在管道中没有任何东西阻碍气流。
—	DECREASE (*v.*) 减少，降低 (*v.*)	A CLOGGED DUCT WILL DECREASE AIRFLOW. 堵塞的管道将降低气流。	A clogged duct will hinder airflow. 堵塞的管道将阻碍气流。
—	CLOGGED (*adj.*) 堵塞的 (*adj.*)	MAKE SURE THAT THE PIPE IS NOT CLOGGED. 确认管道没有堵塞。	Make sure that there is nothing in the pipe to hinder airflow. 确认在管道中没有任何阻碍气流的东西。
hinge (*v.*) 用链连接 (*v.*)	TURN (*v.*) 转动，转弯 (*v.*)	THE PANELS TURN ON TWO NYLON STRAPS. 面板在两根尼龙带上转动。	Panels hinge on two nylon straps. 面板铰接在两根尼龙带上。
HIT (*v.*), (HITS, HIT) 打击 (*v.*)	To touch suddenly and with much force 突然地大力碰触	DO NOT HIT THE CARTRIDGE. 不要击打火药筒。	
hitch (*v.*) 挂接 (*v.*)	CONNECT (*v.*) 连接，联结 (*v.*)	CONNECT THE TOWING ARM TO THE NOSEWHEEL. 连接拖把到前轮部分。	Hitch the towing arm to the nosewheel. 连接拖把到前轮部分。
hoist (*v.*) 升起 (*v.*)	LIFT (*v.*) 升起 (*v.*)	LIFT THE MODULE INTO POSITION. 将组件升起到适当位置。	Hoist the module into position. 将组件升起到适当位置。
hold (*n.*) 保持 (*n.*)	HOLD (*v.*) 持有 (*v.*)	MAKE SURE THAT YOU HOLD THE ROD TIGHTLY. 确保紧紧地握住杆。	Make sure that you have a tight hold on the rod. 确保紧紧握住杆。
HOLD (*v.*), (HOLDS, HELD) 持有 (*v.*)	To continue to have in the hand or grip 用手持续抓住	HOLD THE ROD TIGHTLY. 紧紧握住杆。	—

Word (part of speech) 单词(词性)	Approved meaning/ Alternatives 已批准含义 / 可替换单词	Approved example 批准示例	Not approved 未批准
—	To continue to have in a specified location, position, or condition 在指定的地点、位置或条件下继续	HOLD THE AIRSPEED INDICATION AT THE SAME VALUE FOR 2 MINUTES. 保持相同的空速指示 2 分钟。	
hold back (*v.*) 阻止，抑制 (*v.*)	PREVENT (*v.*) 阻止 (*v.*)	AT FULL THRUST, THE BRAKES MUST PREVENT MOVEMENT OF THE AIRCRAFT. 在全推力情况下，刹车能够防止飞机的移动。	At full thrust, the brakes must hold the aircraft back. 在全推力情况下，刹车能够防止飞机的移动。
	L		
light (*v.*) 点亮 (*v.*)	COME ON 过来，开始	MAKE SURE THAT THE FLUID INDICATOR LIGHT COMES ON. 确认油液指示灯亮了。	Ensure that the fluid indicator light lights. 确保油液指示灯点亮。
LIGHTING (*n.*) 照明 (*n.*)	That which gives light to 使……光亮	A DIMMER UNIT CONTROLS THE PANEL LIGHTING. 一个亮度调节器控制面板照明。	—
LIGHTLY (*adv.*) 轻轻地 (*adv.*)	In a light manner 一种比较轻的方式	LIGHTLY RUB THE DEFECTIVE AREA WITH WET ABRASIVE PAPER. 用湿砂纸轻轻地擦拭有缺陷的区域。	—
LIMIT (*n.*) 限制 (*n.*)	A specified maximum or minimum quantity, number, time, or distance 规定的最大或最小数量、数字、时间或距离	IF THE CLEARANCES ARE NOT-IN THE LIMITS GIVEN IN FIG. 4, REFER TO REPAIR SCHEME No. 2. 如果间隙不在图 4 所示的范围内，请参考修理方案 2。	—
limitation (*n.*) 限制 (*n.*)	LIMIT (*n.*) 限制 (*n.*)	WHEN YOU LIFT THE AIRCRAFT ON JACKS, KEEP THE CENTER OF GRAVITY BETWEEN THESE LIMITS. 当用千斤顶顶升飞机时，保持其重心在限制之间。	Observe these center of gravity limitations when you jack the aircraft. 当顶升飞机时注意这些重心限制。
limited (*adj.*) 有限的 (*adj.*)	SMALL (*adj.*) 小的，少的 (*adj.*)	THERE IS ONLY A SMALL NUMBER OF REPAIRS THAT YOU CAN DO. 只能进行少量的维修。	There is only a limited number of repairs you can do. 只能进行有限的维修。
LINEAR (*adj.*) 线性的 (*adj.*)	In a straight line 在一条直线上	MAKE SURE THAT THE RESULTS ARE LINEAR. 确认结果是线性的。	—
LINEARLY (*adv.*) 线性地 (*adv.*)	In a straight line 在一条直线上	INCREASE THE PRESSURE LINEARLY. 线性地增加压力。	—

续表

Word (part of speech) 单词（词性）	Approved meaning/ Alternatives 已批准含义 / 可替换单词	Approved example 批准示例	Not approved 未批准
link (*v.*) 连接 (*v.*)	CONNECT (*v.*) 连接 (*v.*)	CONNECT THE CABLES. 连接导线（钢索）。	Link the cables. 连接导线（钢索）。
—	ATTACH (*v.*) 连接 (*v.*)	ATTACH THE HOIST TO THE POWER UNIT. 将吊车连接到动力装置上。	Link the hoist to the power unit. 将吊车连接到动力装置上。
LIQUID (*adj.*) 液体的 (*adj.*)	That has the properties of a liquid 具有液体的性质	DURING SERVICING, LIQUID OXYGEN FLOWS THROUGH THE STABILIZING CONTAINER. 在勤务过程中，液态氧会流过稳定的容器。	—
LIQUID (*n.*) 液体 (*n.*)	A material that is not a gas or a solid 不是气体或固体的物质	THE CONVERTER CHANGES THE LIQUID INTO A GAS. 转换器把液体变成气体。	—
list (*v.*) 列出 (*v.*)	RECORD (*v.*) 记录 (*v.*)	RECORD THE TEST RESULTS. 记录测试结果。	List the results of the test. 列出测试结果。
—	SHOW (*v.*) 显示 (*v.*)	TABLE 1 SHOWS ALTERNATIVE ADHESIVES THAT YOU CAN USE. 表 1 显示了可以使用的替代黏合剂。	Table 1 lists alternative adhesives that you can use. 表 1 列出了可以使用的替代黏合剂。
—	LIST (TN) 清单 (TN)	MAKE A LIST OF THE PART NUMBERS IN NUMERICAL SEQUENCE. 按数字顺序列出零件编号清单。	List the part numbers in numerical order. 按数字顺序列出零件编号。
LISTEN(*v.*) （LISTENS LISTENED） 听 (*v.*)	To use your ears to hear or find 使用耳朵去听或去发现	LISTEN FOR THE SIGNAL. 收听信号。	—
little (*adj.*) 小的 (*adj.*)	SMALL (*adj.*) 小的 (*adj.*)	THE DIAMETER OF THE TUBE IS TOO SMALL. 管子的直径太小了。	The diameter of the tube is too little. 管子的直径太小了。
—	SHORT (*adj.*) 短的 (*adj.*)	INSTALL A SHORT LENGTH OF NEW TUBE. 安装一段短的新管子。	Install a little length of new tube. 安装一段短的新管子。
—	NOT SUFFICIENT 不足	IF THE PRESSURE IS NOT SUFFICIENT, THE TEST WILL STOP. 如果压力不够，测试就会停止。	If the pressure is too little, the test will stop. 如果压力太小，测试将会停止。

Word (part of speech) 单词 (词性)	Approved meaning/ Alternatives 已批准含义 / 可替换单词	Approved example 批准示例	Not approved 未批准
little (a little) (*adj.*) 少量的 (*adj.*)	SMALL QUANTITY 少量	ADD A SMALL QUANTITY OF DISINFECTANT TO THE SOLUTION. 在溶液中加入少量消毒剂。	Add a little disinfectant to the solution. 在溶液中加入少量消毒剂。
little(a little) (*adv.*) 少量地 (*adv.*)	SMALL (*adv.*) 小地、少地 (*adv.*)	AFTER YOU REMOVE THE NUT, THE RIB CAN MOVE DOWN A SMALL DISTANCE. 拆下螺母后，翼肋可能会移动一小段的距离。	After you remove the nut, the rib can slip down a little. 拆下螺母后，翼肋可能会下滑一点。
LIVE (*adj.*) 真实的 (*adj.*)	That includes explosive material 包括爆炸物质	SOME MAINTENANCE TASKS ARE NOT PERMITTED ON AIRCRAFT THAT HAVE LIVE AMMUNITION. 有些维护工作是不允许在有实弹的飞机上进行的。	—
O			
overhaul (*v.*) 检修 (*v.*)	OVERHAUL (*n.*) 大修 (*n.*)	DO AN OVERHAUL OF THE No. 2 ENGINE. 对 2 号发动机进行大修。	Overhaul the No. 2 engine. 对 2 号发动机进行大修。
overheat (*v.*) 超温 (*v.*)	TOO HOT 过热	IF YOU PRESSURIZE THE OXYGEN BOTTLE QUICKLY, IT WILL BECOME TOO HOT. 如果氧气瓶加压过快，将会过热。	Rapid charging overheats the oxygen bottle. 快速的充氧会使氧气瓶过热。
OVERLAP (*n.*) 重叠 (*n.*)	The area in which a part of one surface is on a part of a second surface 一个表面的一部分在第二个表面的一部分上的区域	REPLACE THE PART IF THE OVERLAP IS MORE THAN 0.01 mm. 如果重叠超过 0.01 毫米，则更换零件。	—
overlap (*v.*) 重叠 (*v.*)	OVERLAP (*n.*) 重叠 (*n.*)	MAKE AN OVERLAP OF 10 mm. 做一个 10 毫米的重叠。	Overlap the surfaces by 10 mm. 将表面重叠 10 毫米。
OVERRIDE (*v.*), （OVERRIDES, OVERRODE, OVERRIDDEN ） 超控 (*v.*)	To prevent the automatic operation of a part or system 阻止部件或系统自动运行	MANUALLY OVERRIDE THE START SEQUENCE. 人工超控启动顺序。	—
overtighten (*v.*) 过紧 (*v.*)	TIGHTEN TOO MUCH 太紧	DO NOT TIGHTEN THE FITTINGS TOO MUCH. 不要拧接头太紧。	Do not overtighten the fittings. 不要拧接头过紧。

Word (part of speech) 单词（词性）	Approved meaning/ Alternatives 已批准含义 / 可替换单词	Approved example 批准示例	Not approved 未批准
—	TOO TIGHT 太紧	DO NOT MAKE THE FITTINGS TOO TIGHT. 不要拧接头太紧。	Do not overtighten the fittings. 不要拧接头过紧。
—	TOO TIGHTLY 太紧	DO NOT INSTALL THE FITTINGS TOO TIGHTLY. 安装接头不要太紧。	Do not overtighten the fittings. 不要拧接头过紧。
		P	
pack (*v.*) 包装 (*v.*)	PUT (*v.*) 放 (*v.*)	PUT THE ASSEMBLY INTO THE BOX. 把组件放进盒子里。	Pack the assembly into the box. 把组件放进盒子里。
—	FILL (*v.*) 加注，装满 (*v.*)	FILL THE GROOVE WITH GREASE. 在凹槽中加注润滑脂。	Pack grease into the groove. 加注润滑脂到凹槽中。
PAINT (*v.*), （PAINTS, PAINTED） 涂漆 (*v.*)	To apply paint to something 给某物上漆	PAINT ALL THE SURFACES. 把所有的表面都涂上油漆。	—
PAIR (*n.*) 一对 (*n.*)	Two objects that are the same or almost the same, and/or that you use together 两个相同或几乎相同的对象，和 / 或一起使用的对象	MEASURE THE DISTANCE BETWEEN EACH PAIR OF AXLES. 测量每对轴之间的距离。	—
PARALLEL (*adj.*) 平行 (*adj.*)	Along lines that stay a constant distance apart at all points 在所有点之间保持恒定距离的直线	MAKE SURE THAT THE TURNBUCKLE IS PARALLEL TO THE AXIS OF THE AIR OUTLET. 确保螺丝扣与排气口的轴线平行。	—
PARK (*v.*), （PARKS, PARKED） 停放 (*v.*)	To stop a vehicle and to let it stay in one position on the ground 停止交通工具并使其停留在地面上的一个位置	PARK THE CAR. 停车。	—
PART (*n.*) 部分 (*n.*)	1. A constituent of a machine or other equipment 机器或其他设备的组成部分	REPLACE THE DAMAGED PARTS. 更换损坏的零件。	—
	2. A piece or section of a whole 整体的一部分	REFER TO PART 2 FOR THE APPLICABLE PROCEDURE. 有关适用程序，请参阅第 2 部分。	—

Word (part of speech) 单词 (词性)	Approved meaning/ Alternatives 已批准含义 / 可替换单词	Approved example 批准示例	Not approved 未批准
part (*v.*) 分开，分离 (*v.*)	DISCONNECT (*v.*) 断开 (*v.*)	DISCONNECT THE DUCTING. 断开管道。	Part the ducting. 分离管道。
partial (*adj.*) 部分的 (*adj.*)	NOT FULLY 不是全部	IF THE FLAPS DO NOT FULLY EXTEND, DO THE TEST AGAIN. 如果襟翼没有全伸出，请再次进行测试。	If only partial extension of the flap occurs, do the test again. 如果襟翼没有全伸出，请再次进行测试。
partially (*adv.*) 部分地 (*adv.*)	NOT FULLY 不是全部	IF THE FLAPS DO NOT FULLY EXTEND, DO THE TEST AGAIN. 如果襟翼没有全伸出，请再次进行测试。	If flaps only partially extend, do the test again. 如果襟翼仅有部分伸出，请再次进行测试。
PARTICLE (*n.*) 颗粒 (*n.*)	A very small piece of material 一小块材料	IF YOU FIND METAL PARTICLES IN THE PUMP, FIND THEIR SOURCE AND REPAIR THE DEFECTIVE PART. 如果在泵中发现金属屑，请找到其来源并修理有故障的部件。	—
particular (*adj.*) 特别的 (*adj.*)	APPLICABLE (*adj.*) 适用 (*adj.*)	THIS PROCEDURE IS ONLY APPLICABLE TO TYPE A PARTS. 本程序仅适用于 A 型零件 (部件)。	This procedure is particular to type A parts. 本程序仅适用于 A 型零件 (部件)。
		R	
reference (*n.*) 参考 (*n.*)	REFER (*v.*) 参考 (*v.*)	REFER TO CHAPTER 20 FOR THE STANDARD TORQUE VALUES. 有关标准力矩值，请参考第 20 章。	Reference is made to Chapter 20 for standard torque values. 有关标准力矩值的参考在第 20 章。
referenced (*adj.*) 参考的 (*adj.*)	GIVEN (*adj.*) 给出的 (*adj.*)	THIS POINT IS GIVEN ON THE GRAPH. 这个点在图上是已给出的。	This point is referenced on the graph. 这个点在图上是有参考的。
refill (*v.*) 再装满 (*v.*)	FILL (*v.*) 充满 (*v.*)	FILL THE CONTAINER AGAIN. 再把容器装满。	Refill the container. 再注满容器。
refit (*v.*) 整修 (*v.*)	INSTALL (*v.*) 安装 (*v.*)	INSTALL THE LINKAGE AGAIN. 再次安装连杆机构。	Refit the linkage. 整修连杆机构。

Word (part of speech) 单词（词性）	Approved meaning/ Alternatives 已批准含义 / 可替换单词	Approved example 批准示例	Not approved 未批准
—	REPAIR (*v.*) 修复 (*v.*)	ON THE SUBMARINE, REPAIR ALL DAMAGE AND REPLACE ALL WORN PARTS. 在潜艇上，修理所有损坏的部件并更换所有磨损的部件。	Refit the submarine. 整修这艘潜艇。
reflect (*v.*) 反射 (*v.*)	REFLECTION (*n.*) 反射 (*n.*)	THE RETICLE IMAGE MAKES A REFLECTION ON THE BOTTOM SURFACE OF THE GLASS. 交叉线图像在玻璃的底面产生反射。	The reticle image reflects on the bottom surface of the glass. 交叉线图像反射在玻璃的底面。
REFLECTION (*n.*) 反射 (*n.*)	Something that occurs when energy comes against a surface which sends it back 当能量到达一个表面，又被送回	A CLEAN SURFACE GIVES A BETTER REFLECTION. 干净的表面反射效果更好。	—
REFUEL (*v.*), REFUELS, REFUELEP, REFUELED 加油 (*v.*)	To supply with fuel 供应燃油	REFUEL THE AIRCRAFT. 给飞机加油。	—
register (*v.*) 登记 (*v.*)	SHOW (*v.*) 显示 (*v.*)	ADJUST THE SET +40 CONTROL UNTIL THE POINTER SHOWS +40. 调节设置 +40 控制，直到指针显示 +40。	Adjust the SET +40 control until the pointer registers +40. 调整设置 +40 控制，直到指针显示 +40。
regrease (*v.*) 重新添加润滑脂 (*v.*)	APPLY (*v.*) 涂，敷 (*v.*)	APPLY GREASE TO THE ROD AGAIN. 再给杆上涂润滑脂。	Regrease the rod. 再给杆润滑。
—	MORE (*adj.*) 更多的 (*adj.*)	PUT MORE GREASE ON THE JOINT UNTIL YOU CAN MOVE IT. 在接头处多涂些润滑脂，直到能活动为止。	Regrease the joint until you can move it. 再润滑接头，直到可以活动它。
REGULAR (*adj.*) 定期的 (*adj.*)	At specified or equal intervals 以指定的或相等的时间间隔	THE COMPUTER GIVES REGULAR INPUTS TO THE CONTROL SYSTEM. 计算机定期向控制系统输入数据。	—
REGULARLY (*adv.*) 定期地 (*adv.*)	In a regular manner 以通常、有规则的方式	IF THE FAILURE OCCURS REGULARLY, DO A SYSTEM TEST. 如果故障时常发生，需做系统测试。	—

Word (part of speech) 单词（词性）	Approved meaning/ Alternatives 已批准含义 / 可替换单词	Approved example 批准示例	Not approved 未批准
regulate (*v.*) 调节，控制 (*v.*)	CONTROL (*v.*) 控制 (*v.*)	CONTROL THE ELECTRICAL CURRENT. 控制电流。	Regulate the electrical current. 调节电流。
	ADJUST (*v.*) 调整 (*v.*)	ADJUST THE TIRE PRESSURE AS NECESSARY. 按需调整胎压。	Regulate the tire pressure as necessary. 根据需要调节轮胎压力。
regulation (*n.*) 调节 (*n.*)	ADJUSTMENT (*n.*) 调整 (*n.*)	TEMPERATURE ADJUSTMENT IS AUTOMATIC. 温度调整是自动的。	Regulation of temperature is automatic. 温度调节是自动的。
—	CONTROL (*n.*) 控制 (*n.*)	TEMPERATURE CONTROL IS AUTOMATIC. 温度控制是自动的。	Regulation of temperature is automatic. 温度调节是自动的。
—	CONTROL (*v.*) 控制 (*v.*)	A SENSOR CONTROLS THE TEMPERATURE IN THE COMPARTMENT. 一个传感器控制着舱内的温度。	Regulation of temperature in the compartment is effected by a sensor. 舱室的温度调节是由一个传感器来实现的。
reinflate (*v.*) 再膨胀 (*v.*)	INFLATE (*v.*) 膨胀 (*v.*)	INFLATE THE TIRE AGAIN. 再给轮胎充气。	Reinflate the tire. 再给轮胎充气。
reinforce (*v.*) 加强 (*v.*)	MAKE...STRONGER 使……更强大	DOUBLERS MAKE THE JOINT STRONGER. 双接头使连接更稳固。	Doublers reinforce the joint. 双倍的加强接头。
reinstall (*v.*) 重新安装 (*v.*)	INSTALL (*v.*) 安装 (*v.*)	INSTALL THE COVER AGAIN AFTER YOU ADJUST THE UNIT. 校准组件后再安装盖板。	Reinstall the cover after you adjust the unit. 调整完组件后，重新安装盖板。
reinstallation (*n.*) 重新安装 (*n.*)	INSTALLATION (*n.*) 安装 (*n.*)	ONLY APPROVED PERSONNEL CAN DO THE INSTALLATION PROCEDURE. 只有经批准的人员才能执行安装工作。	This reinstallation must be performed by qualified personnel. 这种重新安装必须由授权人员执行。
REJECT (*v.*), (REJECTS, REJECTED) 拒绝 (*v.*)	To make a decision that something is unsatisfactory 做出不能令人满意的决定	REJECT THE PARTS THAT ARE DAMAGED. 拒绝接收损坏的零件。	—

续表

Word (part of speech) 单词（词性）	Approved meaning/ Alternatives 已批准含义 / 可替换单词	Approved example 批准示例	Not approved 未批准
		S	
seepage (*n.*) 渗漏出 (*n.*)	LEAKAGE (*n.*) 渗漏 (*n.*)	CLEAN THE AREAS WHERE THERE IS HYDRAULIC FLUID LEAKAGE. 清洗液压油渗漏区域。	Clean the areas where there is hydraulic fluid seepage. 清洗有液压油渗漏的区域。
seized (*adj.*) 咬住 (*adj.*)	CATCH (*v.*) 抓住 (*v.*)	IF THE CONTROL CABLE IS CAUGHT IN THE PULLEY, RELEASE THE CABLE TENSION. 如果控制钢索卡在滑轮中，松开钢索张力。	If control cable is seized in the pulley, release the cable tension. 如果控制钢索卡在滑轮中，松开钢索张力。
—	MOVE (*v.*) 移动 (*v.*)	IF YOU CANNOT MOVE THE BOLTS, APPLY SOME PENETRATING OIL. 如果不能拧动螺栓，用一些渗透油（除锈剂）。	If the bolts are seized, apply some penetrating oil. 如果螺栓咬住，就用一些渗透油（除锈剂）。
—	TURN (*v.*) 转动 (*v.*)	IF THE FLAP CONTROL MOTOR CANNOT TURN, USE THE ALTERNATIVE MODE. 如果襟翼控制马达无法转动，则使用备用模式。	If flap control motor is seized, use the alternate mode. 如果襟翼控制马达被咬住，则使用备用模式。
SELECT (*v.*) 选择 (*v.*)	Make a choice 做出选择	SELECT THE HYDRAULIC SYSTEM THAT YOU WILL PRESSURIZE. 选择要增压的液压系统。	—
—	—	SELECT A LANGUAGE FROM THE MENU. 从菜单中选择一种语言。	—
—	NOTE: Do not use this word as a synonym for SET. 注意：不要用这个词作为 SET 的同义词	SET THE SWITCH TO "TEST". 将电门设置到"TEST"位置。	Select switch to TEST. 选择电门到"TEST"位置。
SELECTION (*n.*) 选择 (*n.*)	The action or result of choosing 选择的行动或结果	THE OPERATION OF THE INDICATOR DOES NOT PREVENT THE SELECTION OF SYSTEM 1. 指示器的操作不会影响系统 1 的选择。	—

Word (part of speech) 单词（词性）	Approved meaning/ Alternatives 已批准含义 / 可替换单词	Approved example 批准示例	Not approved 未批准
SEMICIRCULAR (*adj.*) 半圆的 (*adj.*)	Has the shape of half a circle 半圆的形状	THE VALVE FLAPS ARE SEMICIRCULAR. 活门挡板是半圆形的。	—
SEND(*v.*)（SENDS,SENT）发送 (*v.*)	To cause to go 致使离开	SEND THE FILTER ELEMENT TO THE OVERHAUL SHOP. 将过滤元件送到检修车间。	—
SENSE (*v.*)（SENSES, SENSED）感觉 (*v.*)	To get an input automatically 自动获取输入	THE TEMPERATURE BULB SENSES THE OUTSIDE AIR TEMPERATURE. 温度探头感受外部空气温度。	—
SENSITIVE (*adj.*) 敏感的 (*adj.*)	That can sense small Changes 能感觉到细微的变化	THE CAPSULE IS SENSITIVE TO PRESSURE CHANGES. 膜盒对压力变化是敏感的。	—
separable (*adj.*) 可分解的 (*adj.*)	DISASSEMBLE (*v.*) 分解 (*v.*)	YOU CAN DISASSEMBLE THIS.UNIT INTO TWO PARTS. 可以把这个组件分解为两部分。	This unit is separable into two parts. 这个组件可分为两部分。
—	DISCONNECT (*v.*) 断开 (*v.*)	YOU CAN DISCONNECT THESE LINE FITTINGS. 可以断开这些管路接头。	These line fittings, are separable. 这些管路接头是可分离的。
separate (*adj.*) 分开的 (*adj.*)	NOT CONNECTED 未连接的	THESE TWO TRACKS ARE NOT CONNECTED TO THE OTHER TWO. 这两条轨道和另外两条轨道是不连接的。	These two tracks are separate from the other two. 这两条轨道和另外两条轨道是分开的。
—	ISOLATED (*v.*) 隔离 (*v.*)	EACH HYDRAULIC SYSTEM IS FULLY ISOLATED. 每个液压系统都是完全隔离的。	All hydraulic systems are completely separate. 所有的液压系统都是完全独立的。
—	NOT ATTACHED 未连接的	THE CAP IS NOT ATTACHED TO THE COUPLING. 盖子未连接到耦合器上。	The cap is separate from the coupling. 盖子与耦合器分离。
separate (*v.*) 分离 (*v.*)	DISCONNECT (*v.*) 断开 (*v.*)	DISCONNECT THE LINE FITTINGS. 断开管路接头。	Separate the line fittings. 将管路接头分开。
SEPARATION (*n.*) 分离 (*n.*)	The action or result of separating 分离的行动或结果	SEPARATION OF THESE PARTS IS NOT EASY. 分离这些部件并不容易。	—

续表

Word (part of speech) 单词 (词性)	Approved meaning/ Alternatives 已批准含义 / 可替换单词	Approved example 批准示例	Not approved 未批准
SEQUENCE (*n.*) 顺序 (*n.*)	The relation of items that follow one after the other in a list or the relation of steps or events that occur one after the other in time 在一个列表中一个接一个地跟随，或在时间上一个接一个地发生的步骤或事件的关系	TIGHTEN THE BOLTS IN THE SEQUENCE THAT IS GIVEN IN FIGURE 3. 按图 3 所示的顺序拧紧螺栓。	—
serious (*adj.*) 严重的 (*adj.*)	IMPORTANT (*adj.*) 重要的 (*adj.*)	VIRUS CONTAMINATION IS AN IMPORTANT PROBLEM. 病毒污染是一个重要的问题。	Virus contamination is a serious problem. 病毒污染是一个严重的问题。
		T	
twice (*adv.*) 两次 (*adv.*)	TWO (*n.*) 两个 (*n.*)	DO THIS PROCEDURE TWO TIMES. 执行该程序两次。	Do this procedure twice. 执行该程序两次。
TWIST (*v.*), (TWISTS, TWISTED, TWISTED) 扭曲 (*v.*)	To use a force that turns something and causes a distortion 使某物转动并引起变形	DO NOT TWIST THE CABLES. 不要扭曲导线束。	—
—	To turn or change shape as a result of torsion 转动或由于扭转而改变形状	IF THE CABLE TWISTS, DISCONNECT THE TWO CONNECTORS. 如果导线束扭曲，则断开两个接头。	—
TYPE (*n.*) 类型 (*n.*)	A specified group 一个指定的组	FIND THE TYPE AND DIMENSIONS OF THE DAMAGE. 查找类型和损伤尺寸。	—
TYPICAL (*adj.*) 典型的 (*adj.*)	That has the important qualities of a group 一个组的重要品质	THIS INSTALLATION PROCEDURE IS TYPICAL FOR THIS TYPE OF FASTENER. 此安装程序是此类紧固件的典型安装程序。	—
		U	
unable (*adj.*) 不能的 (*adj.*)	CANNOT (*v.*) 不能 (*v.*)	IF YOU CANNOT TURN THE PULLEY, MAKE SURE THAT THE PIN IS REMOVED. 如果不能转动皮带轮，确保销子已拆下。	If you are unable to turn the pulley, make sure the pin is removed. 如果无法转动皮带轮，确保已拆下销子。

Word (part of speech) 单词（词性）	Approved meaning/ Alternatives 已批准含义 / 可替换单词	Approved example 批准示例	Not approved 未批准
unauthorized (*adj.*) 未经授权的 (*adj.*)	NOT APPROVED 未批准	IF YOU ARE NOT APPROVED TO DO THIS WORK, DO NOT DO THIS ENGINE TEST. 若未经批准进行该项工作，则不要做该项发动机测试。	If you are unauthorized, do not do this engine test. 如果未经授权，则不要做该项发动机测试。
uncap (*v.*) 打开盖子 (*v.*)	CAP (*n.*) 盖子 (*n.*)	REMOVE THE CAPS FROM THE HOSES. 从软管上拆下盖子。	Uncap the hoses. 打开软管盖子。
unclip (*v.*) 解开 (*v.*)	CLIP (*n.*) 夹子 (*n.*)	REMOVE THE VISOR FROM THE CLIP. 从夹子上拆下遮光板。	Unclip the visor. 拆下遮光板。
uncoil (*v.*) 解开 (*v.*)	UNWIND (*v.*) 解开 (*v.*)	UNWIND THE CABLES CAREFULLY. 小心地解开导线束。	Uncoil the cables carefully. 小心地解开导线束。
uncontaminated (*adj.*) 未污染的 (*adj.*)	CLEAN (*adj.*) 清洁 (*adj.*)	MAKE SURE THAT THE HYDRAULIC FLUID IS CLEAN. 确保液压油清洁。	Make sure that the hydraulic fluid is uncontaminated. 确保液压油未受污染。
	CONTAMINATION (*n.*) 污染 (*n.*)	MAKE SURE THAT THERE IS NO CONTAMINATION IN THE FUEL TANKS. 确保油箱里没有污染物。	Make sure that fuel tanks are uncontaminated. 确保油箱未受污染。
uncouple (*v.*) 断开 (*v.*)	DISCONNECT (*v.*) 断开 (*v.*)	DISCONNECT THE TOW BAR FROM THE VEHICLE. 从车上取下牵引杆。	Uncouple the tow bar from the vehicle. 从车辆上断开牵引杆。
uncovered (*adj.*) 无盖 (*adj.*)	COVER (*n.*) 盖 (*n.*)	DO NOT PUT A COVER ON THE CONTAINER. 不要在容器上盖盖子。	Leave the container uncovered. 把容器盖上。
undamaged (*adj.*) 未损坏 (*adj.*)	NOT DAMAGED 没有损坏	MAKE SURE THAT THE SKIN IS NOT DAMAGED. 确保皮肤没有损伤。	Check that skin is undamaged. 检查皮肤是否完好无损。
UNDEMANDED (*adj.*) 不需要的 (*adj.*)	That occurs without an apparent cause 没有明显的原因	IF YOU GET AN UNDEMANDED MOVEMENT, DO A TEST OF THE SYSTEM. 如果出现不正确的移动，则需做一个系统测试。	—
under (*prep.*) 在下面 (*prep.*)	BELOW (*prep.*) 在下面 (*prep.*)	INSTALL THE CABLE THROUGH THE GUIDE TUBE BELOW THE CABIN FLOOR. 将导线束穿过客舱地板下的导管。	Install the cable through the guide tube under the cabin floor. 将导线束穿过客舱地板下的导管。

Word (part of speech) 单词（词性）	Approved meaning/ Alternatives 已批准含义 / 可替换单词	Approved example 批准示例	Not approved 未批准
—	IN (*prep.*) 在……中 (*prep.*)	THIS CAN OCCUR IN DIFFERENT CONDITIONS. 这可能发生在不同的条件下。	This can occur under different conditions. 这可能在不同的条件下发生。
—	LESS THAN 小于	MAKE SURE THAT THE PRESSURE IS LESS THAN 30 PSI. 确保压力小于 30 PSI。	Make sure the pressure is under 30 psi. 确保压力低于 30 psi。
underneath (*prep.*) 在下面 (*prep.*)	BELOW (*prep.*) 在下面 (*prep.*)	PUT THE CONTAINER BELOW THE DRAIN VALVE. 将容器放在排放活门下面。	Place the container underneath the drain valve. 将容器放在排放活门下面。

Appendix Ⅱ　Words, Phrases & Abbreviations

A

abbreviation	[əˌbrivi'eɪʃ(ə)n]	*n.* 缩写
accelerate	[ək'seləreɪt]	*v.*（使）加快，促进；加速
accessory	[ək'sesərɪ]	*n.* 配件
accumulator	[ə'kjuːmjəleɪtər]	*n.* 蓄压器，储压器
actuator	['æktjuˌeɪtər]	*n.* 作动筒，传动装置
aerodynamically	[ˌeərəʊdaɪ'næmɪkli]	*adv.* 空气动力学地
agency	['eɪdʒənsi]	*n.* 代理行，经销处；政府专门机构
aileron	['eɪləˌrɒn]	*n.* 副翼
aircraft	['eəˌkrɑːft]	*n.* 航空器；飞机
airflow	['eəfləʊ]	*n.* 气流（尤指飞机等产生的）；空气的流动
airworthiness	['erwɜːrðinəs]	*n.* 适航性
alloy	['ælɔɪ]	*n.* 合金
aluminum	[ə'luːmɪnəm]	*n.* 铝
aramid	['ærəmɪd]	*n.* 芳族聚酰胺（高强度阻燃合成纤维，用于制造消防服、防弹衣等）
assembly	[ə'semblɪ]	*n.* 组装；装配 （待装配的）成套部件，零件组合
apron	['eɪprən]	*n.* 停机坪
approximately	[ə'prɒksɪmɪtlɪ]	*adv.* 大概；大约
approval	[ə'pruːvl]	*n.* 批准；同意
approve	[ə'pruːv]	*v.* 批准，同意
authority	[ə'θɔːrəti]	*n.* 当局，权威
autopilot	['ɔːtoʊpaɪlət]	*n.* 自动驾驶仪
automatically	[ˌɔtə'mætɪkli]	*adv.* 自动地
auxiliary	[ɔːg'zɪliəri]	*adj.* 辅助的；备用的
aviation	[ˌeɪvɪ'eɪʃən]	*n.* 航空；飞机制造业

angle of attack		迎角
ALIs (Airworthiness Limitation Instructions)		适航性限制说明
AMECO (Aircraft Maintenance and Engineering Corporation)		北京飞机维修工程公司
AMM (Aircraft Maintenance Manual)		飞机维修手册
ARM (Aircraft Recovery Manual)		飞机恢复手册
ASM (Aircraft Schematics Manual)		飞机原理线路图手册
ASN (Assigned Subject Number)		指定功能号
AWM (Aircraft Wiring Manual)		飞机线路图手册
AWL (Aircraft Wiring List)		飞机线路连接清单

B

beacon	['biːkən]	*n.* 信标
breakdown	['breɪkdaʊn]	*n.* 分类，细目列表
borehole	['bɔːˌhəʊl]	*n.* 钻孔
bulkhead	['bʌlkˌhed]	*n.* 隔板；舱壁
buttock line		纵剖线
bypass valve		通阀

C

cable	['keɪbl]	*n.* 电缆
cap	[kæp]	*n.* 盖；套；罩
category	['kætɪgrɪ]	*n.* 种类，类别
certification	[ˌsɜːtɪfɪ'keɪʃ(ə)n]	*n.* 证明，资质证书
chamber	['tʃeɪmbə(r)]	*n.* 房间，室；腔，膛
channel	['tʃæn(ə)l]	*n.* 频道，电视频道；管道，通道，航道
chord	[kɔːd]	*n.* 翼弦
circumferential	[səˌkʌmfə'renʃəl]	*adj.* 沿边缘的；（尤指）圆周的
civil	['sɪvl]	*adj.* 平民的；民用的
component	[kəm'pəʊnənt]	*n.* 组成部分，成分，部件
combustion	[kəm'bʌstʃən]	*n.* 燃烧
compressor	[kəm'presə(r)]	*n.* 压缩机
configuration	[kənˌfɪgʊ'reɪʃn]	*n.* 结构，构造
corridor	['kɒrɪdɔː(r)]	*n.* 通道
consumption	[kən'sʌmpʃ(ə)n]	*n.* 消费，消耗；食用，引用
cooperation	[kəʊˌɒpə'reɪʃ(ə)n]	*n.* 合作，协作；协助，配合
coordination	[kəʊˌɔːdɪ'neɪʃ(ə)n]	*n.* 协调，配合；身体的协调性
corrosion	[kə'rəʊʒən]	*n.* 腐蚀

crab angle		偏航角
CAAC（Civil Aviation Administration of China）		中国民航总局
CDCCLs（Critical Design Configuration Control Limitations）		重要设计构型控制限制
CG（Center of Gravity）		重心
CML（Consumable Materials Lists）		消耗性材料清单
codes and standards		规范与标准
condition monitoring maintenance		状态监控
corrosion–resistant		耐腐蚀

D

detection	[dɪ'tekʃ(ə)n]	*n.* 察觉，发现；侦破（案件）
dimension	[daɪ'menʃn]	*n.* 尺寸
document	['dɒkjumənt]	*n.* 文件，公文，文献；证件，单据

E

effective	[ɪ'fektɪv]	*adj.* 有效的
elevator	['elɪveɪtə(r)]	*n.*（飞行器的）升降舵
empennage	['empɪnɪdʒ]	*n.* 尾翼，尾部
engine	['endʒɪn]	*n.* 发动机，引擎
exhaust	[ɪg'zɔːst]	*n.* 废气，排气装置
extension	[ɪk'stenʃn]	*n.* 延伸
EASA（European Union Aviation Safety Agency）		欧洲航空安全局
ESPM（Electrical Standard Practices Manual）		电子标准工艺手册
EDP（Engine–Driven Pump）		发动机驱动泵
EMDP（Electric Motor–Driven Pump）		电动马达驱动泵

F

fabric	['fæbrɪk]	*n.* 布料；织物　*adj.* 织物做成的；织物的
failsafe	['feɪlˌseɪf]	*n.* 故障保护，失效保护
fiberglass	['faɪbərˌglæːs]	*n.* 玻璃纤维
filter	['fɪltər]	*n.* 过滤器　*v.* 过滤
fill	[fɪl]	*v.* 注满，充满
flap	[flæp]	*n.* 襟翼
fuselage	['fjuːzəlaːʒ]	*n.* 机身（飞机）
fixed–wing aircraft		固定翼飞机
flight compartment		驾驶舱
flight cycle		飞行周期
front matter		前言

fuel servicing		燃油勤务
FQIS（Fuel Quantity Indicating System）		燃油量指示系统
fuel tank		油箱
FAA（Federal Aviation Administration）		美国联邦航空管理局
FCOM（Flight Crew Operations Manual）		飞机组使用手册

G

graphite	['græfaɪt]	*n.* 石墨
goggle	['gɒgl]	*n.* 护目镜
GAMECO（Guangzhou Aircraft Maintenance Engineering Co.）		广州飞机维修工程公司

H

hangar	['hæŋə(r)]	*n.* 飞机库；飞机棚
hazard	['hæzəd]	*n.* 危险
hose	[həʊz]	*n.* 水管；橡皮软管
hydraulic	[haɪ'drɒlɪk]	*adj.* 液压的
hard time maintenance		定时维修
HMV（Heavy Maintenance Visit）		重度维护检查
honeycomb core		蜂窝芯
horizontal stabilizer		水平安定面
hydraulic fluid		液压油

I

ignition	[ɪg'nɪʃn]	*n.* 点火，点火装置
implement	['ɪmplɪment]	*v.* 执行，贯彻 *n.* 工具，器具
incorporate	[ɪn'kɔːpəˌreɪt]	*vt.* 把……包括在内；将……纳入
infield	['ɪnˌfiːld]	*n.* 内场
indicator	['ɪndɪˌkeɪtə]	*n.* 指示灯
inlet	['ɪnlet]	*n.* (空气、气体或液体进入机器等的) 进口，入口
inspection	[ɪn'spekʃ(ə)n]	*n.* 视察；检查，审视
install	[ɪn'stɔːl]	*v.* 安装，设置
installation	[ˌɪnstə'leɪʃn]	*n.* 安装
intake	['ɪnteɪk]	*n.* 摄入，吸入；入口，进口
interval	['ɪntəvl]	*n.* 间隔
irritate	['ɪrɪˌteɪt]	*vt.* 使过敏；使发炎
ICAO（International Civil Aviation Organization）		国际民用航空组织
IATA（International Air Transport Association）		国际航空运输协会

J

jacking	['dʒækɪŋ]	*n.* 顶托
jet	[dʒet]	*n.* 喷气式

L

laminate	['læmɪnɪt]	*n.* 层压板材
landing gear		起落架；起落装置，着陆装置
leakage	['liːkɪdʒ]	*n.* 渗漏物；泄漏物
lever	['levər]	*n.* 控制杆
leveling	['levəlɪŋ]	*n.* 水准测量；校平
lift	[lɪft]	*n.*（空气的）升力，提升力
load	[loʊd]	*n.* &*v.* 荷载，负担
longitudinal	[ˌlɒndʒɪ'tjuːdɪnl]	*adj.* 纵向的；经度的
lubrication	[ˌluːbrɪ'keɪʃn]	*n.* 润滑；润滑作用
LE（Leading Edge）		前缘

M

magnesium	[mæg'niːzɪəm]	*n.* 镁（化学元素）
maintenance	['meɪntɪnəns]	*n.* 维修；维护
mandatory	['mændətɔːri]	*adj.* 强制性的；法定的
manipulate	[mə'nɪpjəˌleɪt]	*v.* 操纵，控制
manufacture	[ˌmænju'fæktʃə(r)]	*v.*（用机器大量）生产，制造
manufacturer	[ˌmænju'fæktʃərər]	*n.* 制造商；生产商
manual	[ˌmænju']	*n.* 手册　*adj.* 手动的
micro lattice	['maɪkrəʊlætɪs]	*n.* 微格金属，微晶格
military	['mɪlətri]	*adj.* 军事的，军队的　*n.* 军人，军方
modification	[ˌmɑːdɪfɪ'keɪʃn]	*n.* 修改，改变
monocoque	['mɒnəˌkɒk]	*n.* 硬壳式构造
motor	['moʊtər]	*n.* 发动机，马达
MPD（Maintenance Planning Data）		维修计划数据
MMEL（Master Minimum Equipment List）		主最低设备清单

N

nacelle	[næ'sel]	*n.*（飞机的）发动机舱
nanometer	['neɪnəmiːtə]	*n.* 纳米
Nomex		诺梅克斯（一种芳族聚酰胺纤维的商品名）
NTM（Non-destructive Testing Manual）		无损探伤手册
nose landing gear		前起落架

O

outfield ['aʊtfi ːld] *n.* 外场

overhaul [ˌəʊvə'hɔːl] *n.* 大修，彻底检修

On-condition maintenance 视情维修

P

parameter [pə'ræmɪtə(r)] *n.* 界限，范围；参数，变量

pedal ['ped(ə)l] *n.* 脚蹬，踏板

perforation [pɜːfə'reɪʃn] *n.* 穿孔

personnel [ˌpɜːsə'nel] *n.* 职员；人事部门

perpendicular [ˌpɜːpən'dɪkjʊlə] *adj.* 垂直的

pilot ['paɪlət] *n.* 飞行员；驾驶员

pipe [paɪp] *n.* 管子，管道

pitch [pɪtʃ] *v.* 俯仰

precaution [prɪ'kɔːʃn] *n.* 预防措施

preclude [prɪ'kluːd] *v.* 阻止，妨碍

pre-impregnated 预浸渍的

pressure ['preʃə] *n.* 压力

procedure [prə'siːdʒə(r)] *n.* 手续，步骤

pneumatic [njʊ'mætɪk] *adj.* 气动的

pump [pʌmp] *n.* 泵 *v.* 用泵抽送

page block 页面块

PCU（Power Control Unit） 电源控制组件

pitch trim control 俯仰配平控制

power plant 发电厂；动力装置

PMS（Process Material Specification） 工艺材料规范

PP（Practices and Procedures） 实操与程序

primary flight control 主飞行控制

PSEU（Proximity Switch Electronics Unit） 接近电门电子组件

PTU（Power Transfer Unit） 动力转换组件

Q

QRH（Quick Reference Handbook） 快速检查单

quantity ['kwɑːntəti] *n.* 量，数量

R

radius ['reɪdiəs] *n.* 半径

rear [rɪə(r)] *n.* 后部，背部

refuel	[riː'fjʊəl]	*vt.* 给……（再）加油；给……续燃料 *vi.*（再）加油；续燃料
regular maintenance		定期维修
regulation	[ˌregju'leɪʃn]	*n.* 法规，章程
release	[rɪ'liːs]	*n.* 放出；排放　*vt.* 释放
renovation	[ˌrenə'veɪʃ(ə)n]	*n.* 翻新，整修
reservoir	['rezərvwaːr]	*n.*（液压）油箱
resin	['rezɪn]	*n.* 树脂
respiratory	['respɪrətərɪ]	*adj.* 呼吸（器官）的；影响呼吸（器官）的
revision	[rɪ'vɪʒn]	*n.* 修改；校订版
rib	[rɪb]	*n.* 翼肋
rigid	['rɪdʒɪd]	*adj.* 刚性的；坚硬的；不易弯曲的
rudder	['rʌdə(r)]	*n.* 船舵；飞机方向舵
roll	[roʊl]	*v.* 滚转
regulatory agency		监管机构

S

self-healing	[self'hiːlɪŋ]	*adj.* 自恢复性能的（自行净化的）
semi-monocoque	['semi 'mɒnəˌkɒk]	*adj.* 半硬壳式的
sensor	['sensə(r)]	*n.* 传感器
service	['sɜːvɪs]	*n.* 公共服务系统，公共事业；服务；接待
shutdown	['ʃʌtˌdaʊn]	*n.* 关闭；停止运行
slat	[slæt]	*n.* 缝翼
specification	[ˌspesɪfɪ'keɪʃn]	*n.* 规范，说明书
spar	[spaː]	*n.* 翼梁
spray	[spreɪ]	*n.* 喷雾；水沫
spoiler	['spɔɪlər]	*n.* 扰流板
stability	[stə'bɪləti]	*n.* 稳定（性），稳固（性）；坚定，恒心
stabilizer	['steɪbɪlaɪzə]	*n.*（航空器的）稳定器
standard	['stændəd]	*n.*（品质的）标准，水平，规范；正常的水平，应达到的标准
standby	['stændbaɪ]	*adj.* 备用的
strength-to-weight		强度质量比
stringer	['strɪŋə]	*n.* 纵梁；桁条
sump	[sʌmp]	*n.* 滑油罐
surmount	[sə'maʊnt]	*vt.* 克服；越过

support	[sə'pɔrt]	*n.* &*v.* 支撑，支持
safety harness		安全带
scheduled maintenance		定检
secondary flight control		辅助飞行控制
seat track		座椅轨道
service bulletin		服务通告
shock strut		减震支柱
shimmy damper		减摆器
speed brake		减速板
station line		站位线
SDS（Systems Description Section）		系统说明部分
SFAR（Special Federal Aviation Regulation）		特别航空管理规定
SM（Standard Manual）		标准手册
SRM（Structure Repair Manual）		结构修理手册

T

tank	[tæŋk]	*n.* 槽，箱
taxi	['tæksɪ]	*n.*&*v.* 滑行
titanium	[taɪ'teɪnɪəm]	*n.* 钛（化学元素）
thrust	[θrʌst]	*n.* 动力，推力
tolerance	['tɒlərəns]	*n.* 容许偏差
tricycle landing gear		前三点式起落架
trim	[trɪm]	*n.*&*v.* 配平，配平片
towing	['təʊɪŋ]	*n.* 牵引支架
turbine	['tɜːbaɪn]	*n.* 涡轮机，汽轮机
turbofan	['tɜːbəʊˌfæn]	*n.* 涡轮风扇发动机
turbojet	['tɜːbəʊdʒet]	*n.* 涡轮喷气飞机
thrust reverser		反推装置
TE（Trailing Edge）		后缘
transmittal letter		手册发送说明
TSM（Trouble Shooting Manual）		故障分析手册

U

undercarriage	['ʌndərˌkerɪdʒ]	*n.* 着陆装置
unserviceable	[ʌn'sɜːvɪsəbl]	*adj.* 不适用的
utilize	['juːtəlaɪz]	*vt.* 使用；利用

V

validity	[vəˈlɪdəti]	*n.* 合法性；有效性
valve	[vælv]	*n.* 阀门，活门
velocity	[vəˈlɒsəti]	*n.* 速度，速率；高速，快速
ventilation	[ˌventɪˈleɪʃən]	*n.* 空气流通；通风
versus（VS）	[ˈvɜːsəs]	*prep.* 对抗，与……
vertical stabilizer		垂直安定面
visual inspection		目测；目检；外观检验

W

water line		水线
wing spar		翼梁
workload	[ˈwɜrkˌloʊd]	*n.* 工作量

Y

yaw	[jɔ]	*v.* 偏航
yaw control		偏航控制

References

[1] GAMA Specification No.1：Specification for Pilot's Operating Handbook.

[2] GAMA Specification No.2：Specification for Manufacturers Maintenance Data.

[3] GAMA Specification No.7：Specification for Continuing Airworthiness Program（CAP）.

[4] Specification ASD-STE100：Simplified Technical English.

[5] FAA：https://www.faa.gov/.

[6] CFR：https://www.ecfr.gov/.

[7] 黄德先. 民航翻译理论与实务 [M]. 北京：中国民航出版社，2022.

[8] 李向新，邓岚. 飞机机械维修技能基础 [M]. 北京：北京理工大学出版社，2021.

[9] 陈华妮，黄大勇. 民航飞行特情实录英语通话 [M]. 北京：中国民航出版社，2020.

[10] 宋静波，李佳丽. 波音 737 NG 飞机系统 [M]. 北京：航空工业出版社，2016.